INDIAN CARTOGRAPHY

A Historical Perspective

INDIAN CARTOGRAPHY

A Historical Perspective

P.L. MADAN

MANOHAR
2001

First published 1997
Reprinted 2001
ISBN 81-7304-177-6

Published by Ajay Kumar Jain for
MANOHAR PUBLISHERS & DISTRIBUTORS
4753/23 Ansari Road, Daryaganj, New Delhi 110 002
Laser typeset by Pardeep Kumar Goel for Aditya Prakashan,
F-14/65, Model Town II, Delhi 110 009 and
printed at Rajkamal Electric Press, G.T. Karnal Road, Delhi 110 033

Contents

List of Maps

List of Maps

Preface

It was on 14 July 1955 on being posted to the newly created Map Section, when I joined the National Archives of India, that I had my first introduction to the cartographic records of the Government of India that were lodged there. That first contact with the beautiful, hand drawn, original maps became a life-long obsession.

During a period of a little over three decades of my service in the National Archives I was directly associated with the cartographic treasures—maps, and allied materials like field-books, memoirs etc., along with the correspondence records ranging between 1700 and the early twentieth century. It was while going through these map-records individually that I got curious to know the historical back-ground of these valuable maps. A search was made to locate such literature.

However, apart from a few works mostly by European writers, giving *their* interpretation of the growth of the early geography of India, generally in a scattered manner, not a single book giving comprehensive, cohesive and continuous history of the Indian maps could be located. Their were two exceptions however: one book was by Clement Markham—*Memoir of the Indian Surveys,* published in 1773 and 1878; another volume with the same title written by C.E.D. Black, covering the period from 1875 to 1890, was brought out in 1891. But these authors did not have access to the huge correspondence and other map-records of the Survey of India available in India. As a result their books were based mostly on reports etc., that were available in England, and dealt with the early English surveys only briefly.

The only author of the twentieth century who had access to cartographic treasures both in India and abroad, and with whom I have had the honour to discuss various aspects of the Indian cartography many a times, was the late Colonel R.H. Phillimore of the Survey of India. In his five volumed *The Historical Records of the Survey of India* (1945-63), he has recorded in encyclopedic details the activities of the department, and the surveyors and other personalities who were connected with it. Phillimore's book also includes their personal as well as official lives in India and abroad together with covering the professional, and also historical aspects of the evolution of the modern system of survey in India. This admixture was so jumbled up with work-a-day details and extensive quotes from the original texts, etc., that it lost sight of cohesion and continuity. This made the story dull, boring and cumbersome.

Unfortunately, before the fifth volume came out in open market, Phillimore's volumes were frozen from circulation by the Government of India and were treated as 'restricted' publications since nineteen-sixties for reasons best known to the authorities even though the first four volumes were on sale for a long time earlier. These are freely available in many libraries in India and abroad even now, and many scholars must be having their personal copies as well.

More recently another scholar, Susan Gole, has written five books, in which she touched upon some scaterred aspects of the Indian cartography briefly and profusely illustrated them with maps from both Indian and foreign sources which would not, otherwise, have been accessible to many Indian scholars.

The historical maps in the custody of the National Archives of India were generally looked upon as mere 'annexures' to the political events that took place in India, or as archival material, like any other departmental records, lodged there. But, after my interaction with the international community of scholars during the deliberations of the Sixth International Conference on the History of Cartography at London in 1975, the map-records in India began to appear in a different light altogether. It was later found that these map-records were parts of an intrinsic whole and had their own metabolism illustrating practically every aspect of the growth and evolution of the history of Indian cartography since the advent of the English in India till today; only its past was to be explored and brought to light to give it cohesiveness.

During my stay in London, both in 1975 and 1985, covering in all a period of little over six months, I had an opportunity to study the evolutionary history of Indian cartography and examine maps, and also to locate the cartographic literature available in various libraries and museums. The search was continued in Delhi also. I was surprised to discover that two libraries, the Central Secretariat Library and Hardayal Municipal Public Library (formerly known as Hardinge Library) contained extremely rich and valuable material on the subject, although mostly by foreign writers on cartography in general, touching the Indian scenario somewhat superficially.

India being an ancient country, was known to early civilisations. To the West India was an enchanting land somewhere in the east, and to the contemporary Chinese it was located in the west; but its exact location and dimentions were a matter of mere conjecture.

Initially, trading communities from both India and the West opened the way, and the Indian land began to take form in the imagination of the contemporary civilised world. This shape found expression in their early maps according to their respective geographical and cosmological understanding of the universe. The land frontiers of India, being mostly the creation of traditions and political compulsions, kept changing from time to time through the ages, but the coastline was being described and depicted by various countries as per their geographical and cosmological understanding.

Since almost all the pioneers in the field of early cartography came from the West, the entire original literature dealing with early maps, most of which are extant, and their publications, are treasured in the museums and libraries there. Hence, the chapters dealing with the pre-1600 history of Indian cartography are based on secondary sources. The original records, including maps of the Survey of India and its controlling Departments/Ministries of the Government of India constitute the subject matter for the post-1600 period.

I am extremely grateful to the authorities in the ICHR, the then Chairman Dr. Irfan Habib and Dr. T.R. Sareen, Dr. Raj Kumar, the Director and Deputy Director respectively, and others for their assistance which enabled me to complete this study.

I am also grateful to my former Directors, particularly Dr. S.A.I. Tirmizi, and the present Director General of the National Archives of India Dr. R.K. Perti, for giving me an opportunity to remain constantly in touch with the cartographic treasures in the Department even when I was not directly concerned with the Map Section.

For the prompt supply of the relevant records and maps and the efficient service in the Research Room of the National Archieves of India, I am grateful to the staff of the Records Section, Map Section, Library, and the Research Room.

Grateful acknowledgements are also made on the individual maps to the authors and publishers for permitting me to use their maps.

New Delhi P.L. MADAN

Abbreviations

Cat.	:	Catalogue
Con.	:	Consultation
Court.	:	Court of Directors
DDn.	:	Dehra Dun
Deptt.	:	Department
DSG.	:	Deputy Surveyor-General
Ed.	:	Editor
Edn.	:	Edition
F.	:	Folio
Genl.	:	General
G.O.I.	:	Government of India
Govt.	:	Government
Mily.	:	Military
N.A.I.	:	National Archives of India
Offg.	:	Officiating
Ori.	:	Origin
Prog.	:	Proceedings
Rep.	:	Report
SG	:	Surveyor-General
SOI	:	Survey of India
SOS	:	Secretary of State for India
Vol.	:	Volume

1. India and the Antiquity

3000 BC TO THE END OF BC ERA

Interaction between the highly developed ancient civilizations of India, Mesopotamia (Iraq) and Egypt is generally believed by the Assyriologists, including Dr. Saycee, to have begun in the third millennium BC.[1] It was around this time that Indians had reached Mesopotamia and Egypt.

Archaeologists and historical researchers have discovered that Ur, the Biblical city in the south of ancient Babylonia in modern Iraq about 225 km south of Babylon, was the capital of the Mesopotamian valley under the Sumerian kings of the first Dynasty. In another nearby site of excavations at Al'Ubaid, a suburb of Ur, a few personal inscriptions have confirmed that the legendary ruler, Sargon of Akkad (Agade), actually ruled the country in the twenty-fourth century BC.

The city of Ur, having access by river and canal to the Persian Gulf, was the natural centre of foreign maritime trade. As early as in the reign of Sargon of Akkad, mentioned above, Ur had been a flourishing port and was in direct maritime contact with India.

It is generally believed that Indians had established maritime contacts with China and Africa by the seventh century BC.[2] But definite archaeological and historical evidence indicating possible Indian influence on or connection with South Central Africa is available from the Iron Age (100 BC-AD 1500).[3] There is further evidence that Indian traders had commercial contacts with the present-day Rhodesia, during the first century of the Christian millennium. The Rhodesians traded mostly gold and ivory for manufactured Indian articles. It has now been established that the early Iron Age was experienced much earlier in the interior of Africa, in India and also on the East African coast; and the use of iron reached East Africa from India.

Aryanisation of South India began in about 1000 BC and continued for nearly eight centuries. Even during this period of fusion of the two great cultures, the early maritime contacts of India with the countries to its west and east were maintained.[4]

Whether the people from the Harappan civilization living in the Indus valley, or in the Saurashtra region in the western India or the Dravidians inhabiting the Western Ghats of southern India took the initiative and were the first to reach the West, is a point that needs probing in depth.

Nevertheless, it is indisputable that the Indian land mass must have had a place in any mental or figurative conception of the universe of the early civilizations whether Egyptian, Mesopotamian including Sumerian, Babylonian and Assyrian, Iranian, and even Chinese—all older than that of India or Indus valley or even the Dravadian. India was surely known to the early people, at least of the present-day Middle-East region, by the third millennium BC, or even earlier, though no map of the world, other than that of early Babylonian, on clay tablet, referred to later, is extant now. The articles of Indian origin found at many of these places and dated by archaeologists to be of the period round 3000 BC, surely did not reach there just then. This type of trade relations must have taken quite a long time to develop.

The fact that the Indians could reach Mesopotamia, Egypt, Asia Minor and Africa and could return to India so early in time, implies that in antiquity they had acquired the skill of navigating the Indian Ocean. As we shall see, they were first able to make long coastal voyages and later acquired and mastered the knowledge of the air and ocean currents in the Indian Ocean, along with the seasonal behaviour of these currents and made use of this knowledge in their maritime explorations much before the Westerners could even venture out of their comparatively smaller area of operations in the Persian Gulf, Red Sea, and the Mediterranean Sea.[5]

The archaeological findings cannot always be precisely dated and are, therefore, somewhat controversial. From the historical researches in the western literary sources of the ancient historical era, we find some works more reliable such as *Geographica* of Strabo (first century BC), *Natural History* of Pliny (first century AD), and the *Periplus of the Erythrean Sea* (third century AD), the work of an anonymous Alexandrian merchant.[6] In them, we find some clarity of thought and thus they make the periodisation of events somewhat easier. In the absence of direct reliable written contemporary historical evidence even in the West, we have to rely on these sources which, of course, were published much later than the time of events. The Indian literature, similarly, though not strictly historical in nature, and written in Sanskrit and Pali, like *Sutapitaka* (fifth century BC) and *Baveru Jataka* (fourth century BC), give ample proof of historically and geographically significant events of antiquity.[7] These throw enough evidence connected with early navigation and geographical discoveries, hinting at cartographic activities in the period before these were written. By proper study and interpretation of these literary sources, many of the misconceptions have now been removed. One instance was the discovery and applicability of monsoon winds in the Indian Ocean which the *Periplus* attributed to a Greek, Hippalus. This, as we shall see, was erroneous.

Did the Indian navigators at this time have any maps or charts, however crude in form, to guide them in their long voyages to distant lands? And did they have any maps of the Indian coast as well to identify their own part of the homeland on return? Or was the knowledge of map making non-existent in ancient times in India?

How and when and at what stage in the history of Indian navigation, sea charts and maps, obviously with no graticules in the early stages and with no coordinates of Indian origin, reflecting Indian art of navigation, were introduced, is not precisely known. But, it seems certain that sea charts or maps came later after the 'Heaven Finding Art of Navigation' or the method of finding one's bearings in the open sea with the help of the sun during the day and the heavenly bodies at night, had been mastered.

Although no contemporary antique map or chart, however primitive in form, has been discovered so far, there are references to the existence of such maps or charts of somewhat later dates. There is evidence enough to say that Indian navigators had some kind of pilot books as well during the early Middle Ages to guide them. This is borne out by the fact that the Turkish cartographer, Seidi Ali, had used some of them himself.[8] So did the Portuguese on their earlier voyages in the Indian Ocean. This is proved by the information given in their earlier maps about the countries of the east even before they or anyone else from the west, had visited these places.[9]

If we carefully study the works of Marco Polo, Ira Fauro, Nicolo de Conti and many of the early Portuguese explorers, they all point, however vaguely, to the existence of old indigenous Indian marine charts or maps. Even Arabs and Chinese navigators and cartographers admit to having incorporated in their works the information and experiences gained by the Indian navigators, especially belonging to Chola kingdom of southern India.[10] While narrating the voyage of Vasco da Gama, Jean de Barros had stated that da Gama had seen Arab charts. This part of the narrative about Arab charts of the Indian Ocean did not occur in the earlier accounts by Correa and Constanheda. G.R. Tibbetts has doubted the correctness of Barros' statement as his account contained material that was included much after the events had taken place.[11] But this can be easily explained. In his official capacity, Barros had access to all the documents and reports, etc., lodged with the *Casa Da India* which kept on adding and increasing by fresh accumulations. Barros was a prolific writer and he had his influencial sources of information also. Even the manuscript accounts of other explorers, preserved at the *Casa*, were available to him while writing his *Decades*. He used to verify all his geographical locations, before using them, from written as well as oral accounts of those navigators and pilots who had personally been to that region and had sailed in those waters. He was so thorough in verification of place names, their location, etc., that he even did not hesitate to point out mistakes of Ptolemaic geographers who used to follow Ptolemy blindly and were regarded as authorities on the subject. Apart from the evidence from the European sources, Barros used to procure native accounts of the regions. It has been stated that for his comments about India in his *Decades*, he got the help of Persian, Arabic and Indian scholars to translate the regional manuscripts for him. Also he is once known to have asked a Chinese slave to translate Chinese books. He was in the habit of collecting oral history as well of places, which have no written accounts. By using his information, thus collected, in a judicious manner, he produced his *Decades*.[12] Evidently Barros was adding fresh information 'judiciously' Whenever he got hold of a piece of new information, he used to improve upon or update his earlier narrative and so his intentions, perhaps, could not be doubted.

Further, Tibbetts seems to be of the opinion that there were no early indigenous maps or charts of any littoral country of the Indian Ocean, including India, which could be made use of in navigation. The charts that were seen by the Portuguese, or more precisely, Vasco da Gama, were, according to him, merely literary compositions as were being used by Arabs in manuscript atlases and other geographical narratives of Mustaufi, Hafiz-im-Abru and others. He goes on to give Barros' story that Malemo Cana, a Moor from Gujarat, showed Vasco da Gama a peculiar map of the coastal region of India, the like of which da Gama perhaps had not seen or known about earlier. This map had graticules of meridians and parallels drawn very close to each other. The manner of laying down the bearings in this way, without the bearings of the compass, to which he was accustomed, was not the Portuguese way but that of the Moors. Portuguese maps and sea charts invariably had compassrose and bearings.[13] Perhaps, Barros, by the time he wrote this, did not have with him details about the techniques being used by the early Indian sailors while crossing the high seas or was ignorant of the fact that early Indians had already known and mastered the art of 'parallel sailing' and that the map, probably, was being used as a guide for seafaring. Or perhaps Barros was simply repeating what Vasco da Gama had told him, without his own comments. It looks strange that when he was shown the map by the Gujarati sailor, Vasco da Gama did not ask him about the purpose of the north-south bearings on the map, or the absence of compass bearings, and Barros, therefore, could not know of it. Or, in all probability, being a sailor of great experience himself, da Gama knew the purpose of the bearings and so did not ask Malemo Cana about them. The 'meridians and parallels' could have been the bearings for the 'parallel sailing' or for 'running down the latitude', a method commonly used by Indian sailors in medieval India. During the day, he used to sail by observing the latitude of the noon sun and during the night voyage, he used to observe *Dhruva nakshatra* or Pole star while sailing due north or south till the particular latitude of his destination was reached. At that time, the sailor would sail east or west, as was necessary till the familiar landmark was sighted. Then, by visual sailing, he would enter the required port.[14] Or, at this stage, perhaps, he used to consult some sort of guide books, charts, or maps to identify the land in support of his memory and be doubly sure. The researches of another scholar, more specialized in the historical geography of Gujarat, the region which was in the forefront of early navigation, reveal that the Dravidians of the southern India used to navigate the seas with the help of charts that were covered with close network or graticules cutting each other at right angles and such charts were being used by them in the 'very ancient period'.[15]

Yet another eminent scholar has given the details of these graticules and says that the latitudes or the horizontal lines were drawn by observing the altitude of the Pole star, the Calves of the Little Bear and also the Barrow of the Great Bear from the Horizon. *Isba* or an inch, each equivalent to 1042' 50", was the unit of altitude. Again, each *Isba* was sub-divided into *Zams* or eights. The distance between two parallels or *Zams* was equal to 12' 15".[16]

V.A. Janaki has already stated that at the time of the visit in the thirteenth century by the famous Marco Polo, to the East, sea charts were being used by the Indian pilots. And when the Portuguese saw these charts in the fifteenth century, they also, as perhaps did the earlier foreign navigators, mistook the interval between the parallels on the charts for degrees. The details of these measurements etc. had also been explained in great detail by the Turkish Admiral, Ali Ben Husein in his *Encyclopaedia of the Sea*, published in 1584.[17] There is, therefore, no doubt that Vasco da Gama also must have seen such a chart which has been commented upon by G.R. Tibbetts, as mentioned above.

In point of fact Indians had a very long tradition of passing down the knowledge orally to successive generations; and even when the art of writing was more popular and common, people generally *wrote* on religious and literary aspects of life and philosophy. Writing on historical topics or describing the geography of a region in detail, including travel accounts, was simply not there in ancient and to a lesser degree in medieval times in India. The art of navigation, however, was as old as the history of man itself. The oral tradition is so strong that even today, in the age of science and technology, the knowledge of the 'Heaven Finding Art of Navigation' is still passed on orally to the next generation among many of the fisherfolks of the south coastal India. Whatever precious little still remains of this glorious oral tradition has been recorded for posterity by Prof. B. Arunachalam.[18] That Vasco da Gama had seen such a map as described by Barros, therefore, was no surprise. It could have been that Malemo Cana belonged to a family of navigators, with a rich tradition of sailing with the help of old maps or charts which were, perhaps, the family heritage, and was crossing the Indian Ocean like his forefathers when he came in contact with Vasco da Gama.

However, B. Arunachalam, Prithvish Nag, Lotika Vardarajan, B.N. Goswamy, A.L. Dallapicolla, V.A. Janaki and many others, have unearthed enough evidence to establish that since very early in the history of navigation, Indian sailors could navigate the open seas and had some sort of maps and charts to help them to successfully cross the Indian Ocean at just about the beginning of the Christian era.

India, being an ancient country with rich civilization and culture, was known to other contemporary civilizations in the world since very early in the history of man. That it did not develop tradition of writing treatises or travel itineraries, looks extremely strange. In the absence of historical and geographical literature of its own, we have to look mostly towards the west to know about the contemporary geographical and cartographic information about our own country.

The earliest definite evidence of trade between India and the Phoenicians of the Levant, is of 975 BC. After the decline of the Egyptian power, Phoenicians became the master mariners of the Red Sea. King Solomon of Israel got his ships built for his navy in Ezion-Geber, a port on the Gulf of Aqaba in the Red Sea. Some of his ships were constructed at Eloth also on the banks of the Red Sea.[19] Hiram, Phoenician ruler of Tyre, had an alliance with King Solomon. He was an ambitious man. Jointly they traded in the east with the result that a great deal of wealth was accumulated at Jerusalem.[20] These ocean-going ships of Solomon were called the 'ships of Tarshish'. These, along with the navy of Hiram, sailed in 975 BC, from Ezion-Geber to bring ivory, gold, apes and peacocks from the port of Ophir to decorate the palaces and temple of Solomon. They completed the trip to Ophir and back in about three years. Thus, according to modern researches, so far, the Phoenicians were the earliest connecting link between the Indian and the Mediterranean cultures.[21] The *Baveru Jataka*, a Buddhist work of 400 BC, also confirms this and states that Indian merchants took peacocks by sea to Babylon.[22]

There is an interesting controversy amongst the scholars regarding the location of Ophir. It is significant to note in this connection, that the words *tukkiyyim* and *qophin* in Hebrew for peacock and ape respectively, are the very words even now being used by the local people of Malabar on the west coast of India.[23] This goes to prove that peacocks and apes came originally from India in the 975 BC expedition of Solomon or might have been brought even earlier in the region of Levant by the Indian traders, since India's trade relations with Mesopotamia and the region existed since third millennium BC. There is every possibility that Solomon came to know of the existence of this beautiful bird and the apes while moving in some parts of Mesopotamia and also that these were brought to Mesopotamia from Ophir. This discovery by Solomon might have led to his desire to import these animals for his palaces and temple. The Indian names for these animals were very popular, and were, perhaps, incorporated into Hebrew language either in some remote past or during the time of Solomon. This, however, is an issue for linguists. The entire merchandise imported by Solomon, according to another writer, smacks of India etymologically and economically.[24]

The port of Ophir, from where the animals and other things were imported into Israel must have been located in the region of their normal trade, or at a place which must also have had these items in abundance for export. This criteria applies to India more than to any other region in their neighbourhood. In the context of India, the name of Supara, a port about 37 miles north of Bombay, could be cited.[25] Supara's old name was Sophir. Local Gujarati people spell it as 'Sophir', but while pronouncing it, they omit the initial 'S', and that sounds as 'Ophir'. Again, as stated earlier, the time taken to and from Ophir was three years. The time roughly estimated agrees only with Supara as the destination and not with other ports in south-east Arabia, or the north-east African ports with which Israel was, as well, having trade relations. It is very relevant to mention here that after the discovery and the mastery of the monsoon winds first by the Indians and later by the westerners, time taken for a voyage even from Italy to India via Egypt, including the land journey over Egypt, was between four to six months.[26] All these go to suggest that Supra was the ancient port of Ophir. Incidentally, this settles another hypothesis. Solomon ships completed the voyage in three years. This was because of their inability to use the monsoon winds to sail direct to Ophir and back across the Indian Ocean, coasting all the way. It can, therefore, be safely said that by 975 BC, the period when the naval expedition of Hiram and Solomon to India took place, the westerners were ignorant of the monsoon winds.

Interestingly by even during the fourth century BC, the Greeks did not have a clear idea of the hazards in the Red Sea and the Arabian Sea. This is evident from an incident during that period. Alexander, the Great, had despatched sea expeditions to sail round Arabia, from both the Egyptian and the Mesopotamian sides, but the expeditions failed[27] at Bab-el-Mandeb. There could be many reasons. Possibly they were not properly equipped to face the challenges of the open sea voyage in the Arabian Sea waters and its turbulent sea currents, or because of the design and shape of their ships, they could not cross the treacherous strait of Bab-el-Mandeb, in the southern extremity of the Red Sea, and reach the open waters of the Arabian Sea.[28] Alexander, while retreating from India along the coasting route again could not cross the strait of Bab-el-Mandeb from the Ocean-side. But strange as it may seem, ships belonging to Hiram and Solomon could sail all the way to and from Ophir after crossing the strait of Bab-el-Mandeb in the tenth century BC. Again, in the fourth century BC, the Lagid kings of Egypt crossed this obstacle to reach Socotra. Agatharchides, towards the end of the second century BC, informs us that Arabs from the south had big rafts and small boats of skin, which enabled them to successfully go through the kind of hazards as presented at the strait of Bab-el-Mandeb.[29]

There is a little group of islands, off the Somalia coast, to which the early Indian navigators gave the name—*Dvipa Sukhatra*. This is now called *Socotra*, a Greek translation of the earlier Sanskrit name, meaning 'The Happy Isles'. The early Greek geographer and writer, Agatharchides, further informs us that Socotra had become a trading mart and, it was here that merchants and traders from all parts of the known trading world, including Patala (at the mouth of the Indus), Persia and Carmania, used to gather.[30] His mentioning of only Patala signifies that Agatharchides had not heard of any place farther away including southern India.

Socotra, as its earlier Sanskrit name signifies, was discovered by the Indians much before it was frequented by people from the nearby Somalia or south-west Arabia. It came into prominence, perhaps, as a halting stage in the early period when the Indians were reaching Africa, coasting all the way in the seventh century BC or even earlier.

Eudoxus of Cyzicus, at the end of the second century BC, was the first to undertake the voyage from Alexandria to India by sea bringing with him an Indian pilot found on the Egyptian coast. This information is given in the *Geography* of Strabo of the first century BC. It goes to prove that Indians were the first to undertake the voyage from the east to the west prior to the Hellenistic seafaring of Egyptians.

By the second century AD, the ports of South Arabia and Socotra were very popular as meeting places of two cultures and trade, Egypt on one side and India on the other.[31] Egypt at that time in a way, represented the Mediterranean trade and culture or even the entire western world, at these centres.

In between the third century BC, and the second century AD, the entire scenario changed.

As already mentioned, India had reached Mesopotamia in the third millennium BC and had contacted African coast in the seventh century BC. Till then, Indians were following the 'coasting' mode of navigation. It seems that they took a long time to cross over from the Gulf of Oman to south Arabia, to Somalia coast down to South Central Africa, because of peculiar wind conditions, or perhaps their vessels were not sturdy enough to face the wind challenges. Socotra might have been one of the many halting stations on the way. Later, during the third century BC, the Indians had mastered the monsoon winds and by the second century BC, intercourse with the Ptolemaic rulers of Egypt increased considerably, resulting in regular increased trade. Between 120 and 110 BC, there is evidence of sea trade between India and Egypt. During these trading voyages Socotra played a very prominent part.

All these go to prove that right from the third millennium BC to the middle of the first millennium of the Christian era, Indians were, in a way masters of the sea. They had learnt to use the monsoon winds to their advantage in navigation and were able to go direct across the Indian Ocean to the Persian Gulf, Red Sea and even to South Central Africa. They had inspired, in due course, not only the navigators of littoral countries of the Indian Ocean, but the Iberians and the Greeks as well, to learn the art of using the monsoon winds to guide their ships while negotiating across the waters of the Indian Ocean. This had been admitted, directly and indirectly, by many foreign travellers, navigators and traders alike. To say that the use of monsoon currents was discovered by Hippalus, a Greek, in the first century BC, does not seem to be tenable.

So far I have surveyed the world scenario in general in order to trace and locate early signs, in antiquity, of geographic and cartographic discoveries including navigational aids, in the form of sea charts, etc., so far as these were linked with the subsequent development of Indian cartography, in particular.

Some early centres of geographic and cartographic traditions and practices in the pre-Christian era have been located and these are discussed briefly in chapter 2.

NOTES

1. Hibbert Lectures, 1887, *Origin and Growth of Religion among the Babylonians*—quoted in R.K. Mukherjee, *A History of Indian Shipping*, 2nd edn., 1957, p. 60.
2. Charles Verlinden, 'The Indian Ocean: The Ancient Period and the Middle Ages' in Satish Chandra (ed.), *The Indian Ocean Explorations in History, Commerce & Politics*, New Delhi, 1987, p. 30.
3. Prithvish Nag, 'The Indian Ocean: India and Africa: Historical and Geographical Perspectives' in Satish Chandra (ed), *The Indian Ocean Explorations in History, Commerce & Politics*, New Delhi, 1987, p. 152.
4. S.M. Sastri, *M. Crindle's Ancient India as Described by Ptolemy*, Facsimile Reprint with Introduction, Calcutta, 1927, p. 76.
5. R.A.L.H. Gunawardana, 'Changing Patterns of Navigation in the Indian Ocean and their Impact on Pre-colonial Sri Lanka,' in Satish Chandra (ed.), *The Indian Ocean Explorations in History, Commerce & Politics*, New Delhi, 1987, pp. 59-60.
6. Some writers believe it to be of the first century AD.
7. V.A. Janaki, 'Gujarat as the Arabs Knew it' in *Maharaja Sayajirao University of Baroda, Geography Research Paper Series*, No. 4, Baroda, 1969, p. 13; R.A.L.H. Gunawardana, op. cit., pp. 54-89.
8. B. Arunachalam, 'The Heaven-finding Art in Indian Navigational Traditions and Cartography' in Satish Chandra (ed.), *The Indian Ocean Explorations in History, Commerce & Politics*, New Delhi, 1987, p. 202.
9. Leo Bagrow, *History of Cartography*, English Edn., Revised and Enlarged by R.K. Skelton, London, 1064, p. 207.
10. Satish Chandra (ed.), *The Indian Ocean Exploration in History, Commerce & Politics*, New Delhi, 1987, p. 16.
11. Susan Gole, *Indian Maps and Plans from the earliest times to the advent of European Surveys*, New Delhi, 1989, p. 39.
12. Donald F. Lach, *Asia in the Making of Europe*, in 2 vols., Vol. I, Chicago and London, 1965, p. 191.
13. Susan Gole, op. cit., p. 39, quoted from G.R. Tibbetts, 'Oriental Cartography in the Indian Ocean in pre-Portuguese times' to be published in Brian Harley and David Woodward (eds.), *History of Cartography in*

Traditional Societies, University of Chicago Press.
14. B. Arunachalam, op. cit., p. 192.
15. V.A. Janaki, op. cit., p. 7.
16. D.C. Sircar, *Geography of Ancient and Medieval India,* Varanasi, 1950, p. 250.
17. V.A. Janaki, op. cit., pp. 6-7.
18. B. Arunachalam, op. cit, pp. 191-221.
19. G.F. Hourani, 'Arab Seafaring in the Indian Ocean' in *Ancient and Early Medieval Times,* Princeton, N.J., 1951, p. 8.
20. K.A.N. Sastri, *History of South India from Prehistoric Times to the Fall of Vijayanagar,* Madras, 1995, p. 76.
21. H.G. Rawlison, 'Early Contacts between India and Europe', A.L. Basham (ed.), *A Cultural Heritage of India,* Oxford, 1975, p. 76.
22. V.A. Janaki, op. cit., p. 13.
23. *Encyclopaedia Britannica,* 1967, Vol. 16.
24. G.F. Hourani, op. cit., p. 7.
25. H.G. Rawlison, op. cit., p. 425, *Encyclopaedia Britannica,* 1967, Vol. 16.
26. D.F. Lach, op. cit., Vol. I, p. 13.
27. G.F. Hourani, op. cit., p. 19.
28. Ibid., p. 19.
29. Charles Verlinden, op. cit., p. 31.
30. G.F. Hourani, op. cit., p. 22.
31. Ibid., p. 22.

2. Early Centres of Cartography

EARLIEST TIMES TO THE END OF BC ERA

Indian maritime contacts with Babylonia are, perhaps, three millennium years old. Its capital city, Ur, now in ruins, was a well-planned and beautifully laid out port town. It has lately been greatly recovered by the archaeologists. They have bared open its streets, buildings, palaces and temples including other cultural edifices. With consistent efforts, most of the once rich archives and libraries of ancient Babylonia, have been retrieved. Although it had a very rich tradition of map-making, only a few maps could be salvaged as most of them, with the passing of time, have left no trace. It has been, however, established that Babylonians used clay tablets for cuneiform inscriptions of documents. Similarly, their early maps have also been found impressed on such clay tablets.[1] Unfortunately, the clay tablets used for maps were very small which made the deciphering extremely difficult. Also much of the engraving has been disfigured because of time and weather conditions.

Historical researchers say that Babylonians seemed to regard the world as round, encircled with an ocean, which they used to call the Briny or Bitter Waters with seven islands around it, probably representing, as they believed, the seven zones or climates into which the world was divided. These islands were located outside the inner-circle. These islands seemed to form a bridge to the outer-circle and called it Heavenly Ocean. As per their beliefs, it was to these islands that their ancient gods used to be banished.[2]

On one of the early clay tablets, a diagrammatic representation of the world, with Babylon at its centre, is depicted. This is dated from about 500 to 600 BC. This is the first ever world map extant and is now preserved in the British Museum. It is difficult to interpret the details on this small tablet. All that could be established from this schematically portrayed map of the world is—the river Euphrates is flowing down from the Armenian mountains, which the early Babylonians used to call, Urartu. It is flowing to the centre of the city of Babylon. Here it enters the Persian Gulf.[3] The rest could be vaguely interpreted, keeping in view their general belief of the universe. This is all that the scholars could say and the rest is all mystery. There is no denying the fact that India was well known to the Babylonians and to other people of the region by 500 BC, the date fixed by the present archaeologists, of the map-tablet. King Solomon used to send naval expeditions to India since 975 BC and the map tablet has been dated from 500 BC or more; India, therefore, would have been shown in the world map even if it represented merely a block-scheme of the world in the form of a disc floating on the sea.[4] No other countries, including India, have been identified so far, although the map represents their conception of the whole world.

Egypt, like Babylonia, had a very high civilization in ancient days. Its geographical knowledge was also greatly developed and had a rich tradition of map-making as well. The oldest extant papyrus, generally known as Turin Papyrus, is dated about 1320 BC. It is a sketch of an Egyptian gold mine showing as well, the topography of the country around that area, etc. The name of the site of the gold mine is not given. Rameses II (about 1300 BC) is credited with being the first to attempt to survey the Nile lands.[5]

The Pharaohs of Egypt organised geographical expeditions by sea to explore various countries as early as 1492 BC. Herodotus has described one such voyage. A naval expedition under Pharaoh Nicho (*c.* 596-94 BC) sailed down the Red Sea, circumnavigated Africa, and back to Alexandria by way of the Pillars of Hercules, or what is now called the Straits of Gibraltar.[6] However rich the tradition of map-making there might have been, whatever are extant now, are mostly cadastral drawings and plans; but no geographical map of the world, as understood by them in those days, has come to light.

Before the eighth century BC, during the Homeric age, the people of Greece used many articles in their daily life that were made in India. Most of the people had heard of India but had no idea as to its shape, size and even its exact location. Homer and his contemporaries knew very little beyond Greece, the archipelago, Asia Minor, Egypt, Sicily, and part of Italy.[7] By 600 BC, the Greeks had developed the town of Miletus on the Aegean Sea, as a centre for geographical studies and cosmological speculations. It has been established that the first ever world map, if we exclude the Babylonian world map of

Map 1: *Outline map of the World'* by Hecataeus (*c.* 550-480 BC). (Courtesy: Dr. R.P. Misra, *Fundamentals of Cartography*, Mysore, 1969)

500 BC on clay tablet, was drawn by an Ionian named Anaximander (610-546 BC) although neither Herodotus nor Aristotle, later historians and philosophers are stated to have seen it.[8] He is also said to have invented an instrument that could calculate the latitude of a place.[9]

During the sixth century BC, India which lay far away in the east almost like a misty land, although the Indus valley was annexed to the Persian Empire by Darius Hystaspio in 515 BC, once again caught the imagination of the then ruler.[10] In 509 BC, therefore, Scylax of Caryanda, a Greek officer, was asked to go on a fact-finding mission into this region of the easternmost province of their Empire. He sailed down the Indus and also along the southern coast of Arabia.

Scylax also committed the same mistake as others like Herodotus and many others had done earlier, i.e. he was not clear about the directions, as in his report he had stated that the Indus ran in southeasterly direction, whereas it actually flows down in southwesterly direction. His description of other geographical and topographical features also corroborate the statement. He did not give any information about the region beyond the areas where he had travelled except that it was all a sandy desert. Thus, the first ever eyewitness report, was a mixture of fact and fantasy, and was all that the Westerners were fed with.[11]

The next known traveller and geographer, also a

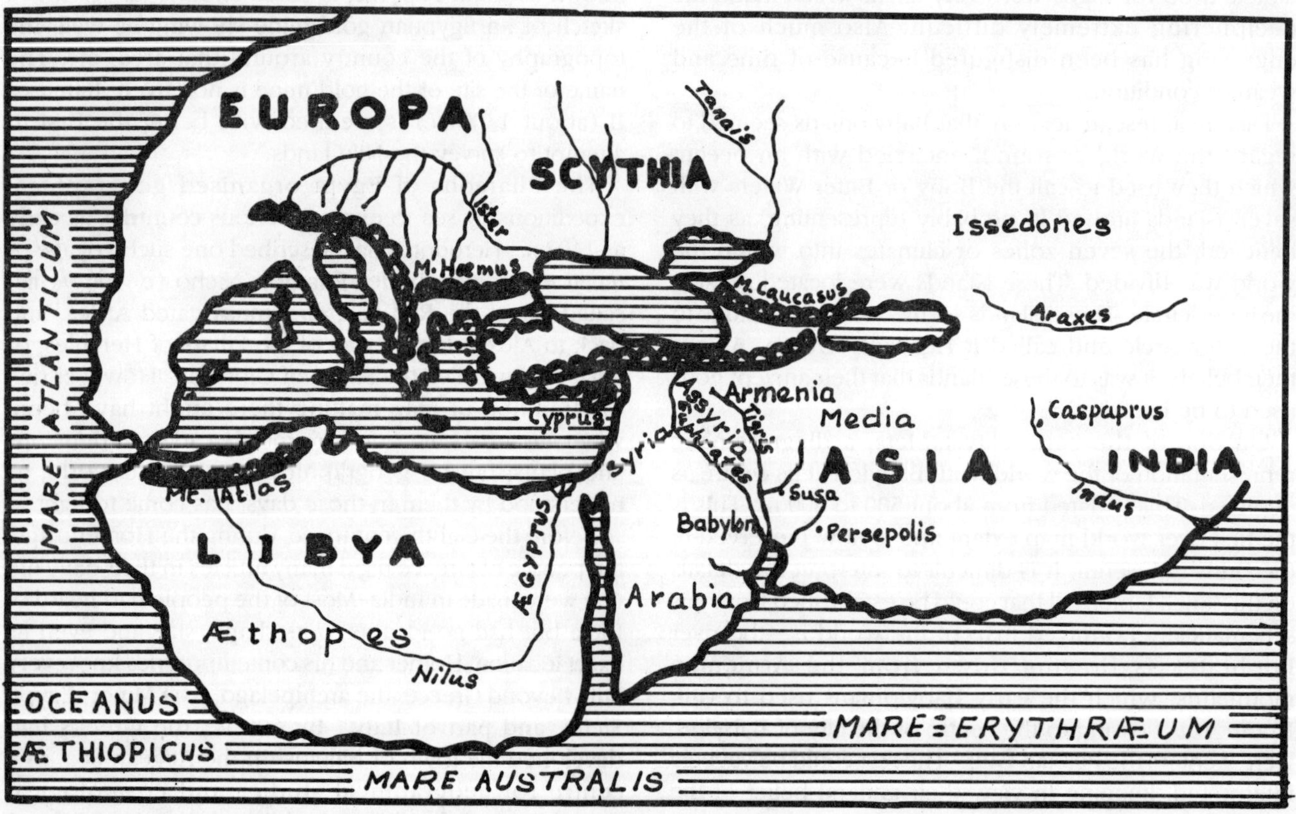

Map 2: *Outline map of the World* by Herodotus (450 BC). (Courtesy: J.M. Dent & Sons Ltd., Publishers, *Atlas of Ancient Classical Geography*, London, 1952).

scholar from Miletus, the great centre of Greek learning, was Hecataeus (*c.* 550-480 BC). His geographical knowledge was mostly confined to two continents only—Europe and Asia, and a little bit of Africa. His work, *Survey of the World*, is not extant now.[12] He is said to have drawn a map as well to accompany his account. But this is now disputed by scholars.[13]

Another historian, Herodotus (484-431 BC), known in the western world as 'the Father of History', was also known to be a great traveller. But the present-day scholars are still perplexed as to where he actually travelled to.[14] Another of his many such puzzles is that he did not agree with the theory of the flat world but what his alternative view was, is not known.[15] All seem to agree that his knowledge of India was very poor and vague. He had no idea even to its extent and exact location. What he was certain about was that it was one of the easternmost province of the Persian Empire or the then known world. His physical description of India was mostly taken from Hecataeus, referred to above, and he also was sure, like Scylax in 509 BC, that Indus ran southeastward.[16]

The *Indica* by Ktesias (398 BC) of Ciudus, a critic of Herodotus, and the royal physician of Persia, is full of 'old wives' tales, and is not to be trusted,[17] although he was the first author to produce a separate work on India. Around 400 BC, he wrote two treatises, *Persica* and *Indica*. In *Indica* he tried to correct Herodotus on Asia including India, but his description of India in it is far less factual than what Herodotus had stated. He even exaggerated the size of India tremendously. He is more factual while describing the routes and distances between India and Ephesus and about the mountain ranges of northern India. But his 'facts are fewer than his fantasies'.[18]

In the fourth century BC, we have the earliest evidence of a scientific approach to cartography from Dicaearchus (350-290 BC), a disciple of Aristotle. He incorporated an orienting line on his world map, running from west to east through Gibraltar to Rhodes extending up to Persia.

Before the rise of Alexander, the mighty Persian Empire was like a great wall separating the Greek world from the New World, i.e. Asia. The image of India known to the Greeks and others, was the one painted by Scylax in his 'eyewitness' account of 509 BC, related earlier, when he travelled down the Indus. It was Alexander's campaign into India from 326-324 BC that helped, to a great extent, to clear the mystery surrounding India, when the Greek civilization came in direct contact with the Indian civilization and culture. The people of both the countries were, till then, remotely conscious of each other's existence,[19] even though the ancient Greeks were using articles made in India.

By the time Alexander died in 323 BC, western scholars had already known much about India and they began sorting out facts from fiction fed to them by earlier descriptions.

A number of scholars and researchers had accompanied Alexander in his campaign into India. These included Kallisthenes of Olynthos who wrote an account of Alexander's Asiatic campaign, and Diognetos and Baton who recorded the measured distances of the army's marches.[20] Unfortunately, the original works of all these are not extant now, but, these were reconstructed by studying the writers of later period like Strabo, Pliny, Ailianos, Arrian, etc., who quoted them at great length.[21] Even the work of Megasthenes, who was later sent as an ambassador to the Court of Chandragupta by Seleukos Nicator and who described the geography in very great detail, has been lost.[22] Those writers who accompanied Alexander had collected enormous data practically of every facet of life in India, including geographical details. Eratosthenes, the great geographer of his times, who was the librarian from 234 to 196 BC of the renowned library at Alexandria which was established by Ptolemies, collected, sorted out and analysed the huge geographical data that was accumulated in Greece and described his conclusions in his famous work *Geographica* in detail.[23]

Eratosthenes was a practical geographer. He calculated the distance between Syene to Alexandria with the help of cadastral maps that were based on actual surveys, and were available in the library, and with these, as the base, he calculated the circumference of the earth, assuming of course, the earth to be a perfect sphere. This was the beginning of the concept of scientific geography based on actual land surveying[24] and mathematical calculations. Behind Eratosthenes's thought was the theory which originated with Thales (600 BC) but propounded by Pythogoras, that the earth was a perfect sphere and lay in the centre of the Universe, which, as we all know now, is not true. Sir E. Bunbury has justly pointed out that Eratosthenes's geography was not *only much nearer the truth than* estimated by Ptolemy three centuries later, but it was actually a better approximation than was arrived at by modern geographers till about two centuries ago.[25]

Eratosthenes was interested in the Greek knowledge about geographical theories. His three books of *Geographica* are not extant but Strabo (63 BC-AD 21),[26] kept his thoughts and theories alive in his works. He took the island of Rhodes as the focal point for calculating the distances. From the data gathered by him, he calculated the shape, size and even the placement of India. It was he who for the first time projected the peninsular character of south India for the proposed map of the world,[27] which Ptolemy (AD 87-150) could not visualize while preparing for his world map; and the mistake initially made by Ptolemy, as we shall see, continued to be believed for centuries.

The tricky problem of representing the spherical earth on a plane surface was also tackled by Eratosthenes. For this, he drew two more parallel, horizontal lines, on his world map, one passing through the Straits of Gibraltar, running through the Caspean Sea, and the other through Egypt and southern India. There were a number of vertical lines to calculate from the island of Rhodes for the representation of the spherical shape on to a plane. His prime meridian was the meridian passing through Alexandria.[28]

About two centuries after Eratosthenes, came Strabo (63 BC-AD 21) the author of six historical-geography books on Asia in Greek; his object in writing a new geography was 'to correct the earlier works in the light of the increase of knowledge' due to the foundation of the mighty Roman Empire.[29] While he incorporated materials derived from his own experience in Asia Minor, he relied mainly upon Homer, Eratosthenes, Polybius and Posidonius for his remarks on the East which were few and also, mostly, self-contradictory. He frequently quoted earlier geographers but while writing on Asia, he seemed to accept Eratosthenes as his guide.[30] In spite of all this, he rightly conceived the difficulties in correctly representing a curved surface on a plane and perceived that a projection must remain erroneous to some extent.[31]

There are many references available that show the existence of maps in ancient Rome. No serious thought was given to the theories of map-making during the establishment of the mighty Roman Empire. Many maps were prepared to illustrate Latin classical texts. We hear of individual world maps as well, such as the one prepared by Marcus Vipsanius Agrippa (63-12 BC). This is titled *Orbis Terrarum* or 'Survey of the World' and is dated 12 BC. It is believed that Roman civilization possessed many other good maps also. In few maps that have survived, roads were marked as straight lines and the distances between any two stages were also given; true directions were not adhered to, which resulted in the distortion of the shape of the country and also the relative positions of various features.[32]

The basic cosmographical concept of Roman cartographers, as of the Greeks, was that the earth was considered a flat disc surrounded by sea. The maps were mostly east-oriented, and the Aegean Sea was the central focal point as in the *Orbis Terrarum*. Though their general concept was very much like that of the Greeks, the Roman map makers of Mela's times and of later period derived very few of other ideas from Hellenistic cartography.[33] India is normally absent from the maps that were produced by Roman map makers during the pre-Christian era.

The Chinese civilization is said to be Asia's oldest civilization. China claims to be the first country in Asia to have prepared and used maps in their administrative matters. Their tradition places the first map in about 2000 BC. The compass is said to be their invention of 1100 BC.

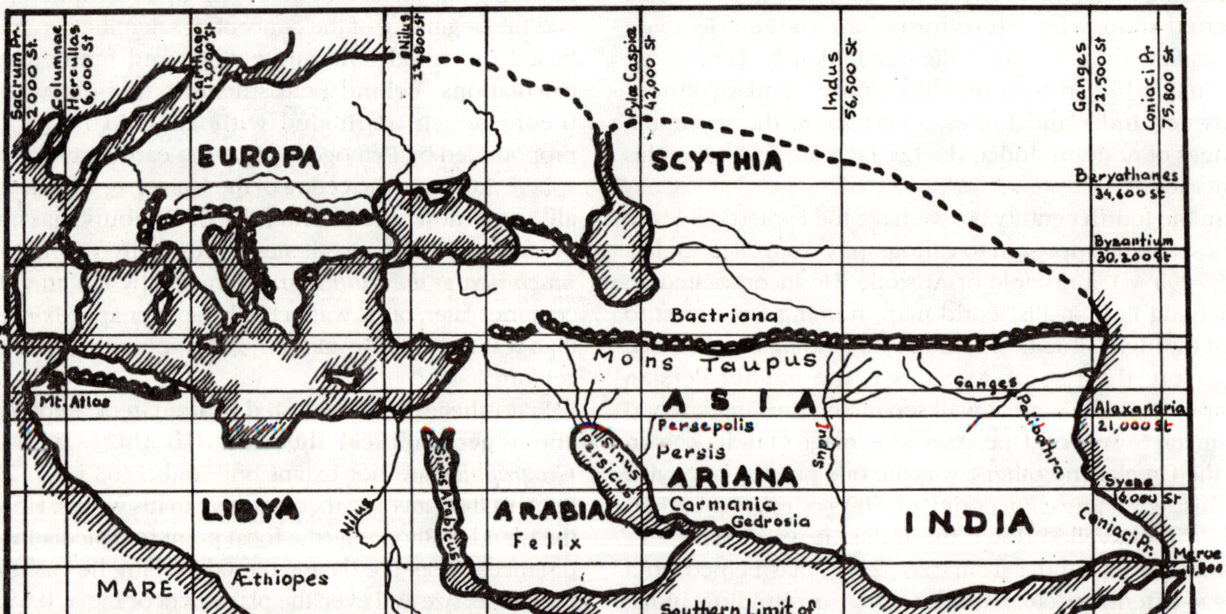

Map 3: *Outline map of the World* by Eratosthenes (220 BC). (Courtesy: J.M. Dent & Sons Ltd., Publishers, *Atlas of Ancient Classical Geography*, London, 1952).

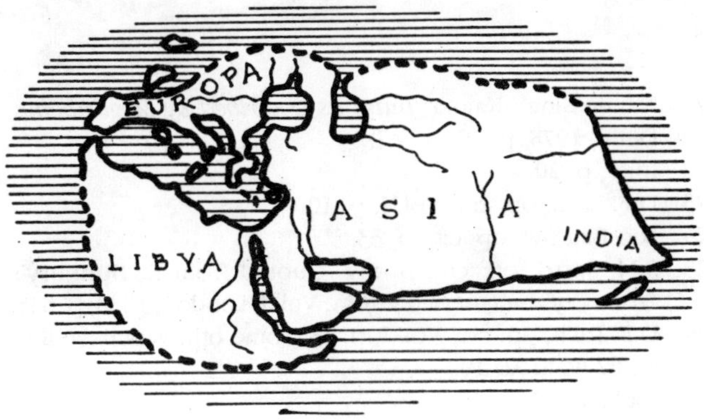

Map 4: *Outline map of the World* by Strabo (AD 18). (Courtesy: J.M. Dent & Sons Ltd., Publishers, *Atlas of Ancient Classical Geography*, London, 1952).

On their maps they used to mark astronomically determined fixed points since very early days.[34]

Until 450 BC the Chinese believed that the earth was square and not round or oval as the Greeks claimed. This concept changed when the Chinese pilgrims visited India and learnt of Indian cosmology. This change could be seen in the Chinese maps of the late fourth century BC. The Indian cosmographic ideas were appreciated in that country as is evident from their maps of the early days.[35]

Modern Chinese scholars claim that the Chinese had crossed the Pacific Ocean to reach the present American continent in about 3000 BC, much much before Columbus 'discovered' it.[36] The Chinese were, no doubt, great navigators and the use of the sea charts was also quite popular since very early days along with the administrative maps. Marco Polo, in the thirteenth century had corroborated that Chinese sailors were familiar with distant Asian ports as well.[37]

Japanese cartography developed quite independently, without outside influence in the early stages of its development. It matured in about the seventh century AD.

Interaction between India, China and South East Asia is said to have begun since the first millennium BC. People of this region were great navigators and like Indians, their forefathers had also crossed the Indian Ocean much earlier in their history of navigation and seafaring. Despite this, no early sea charts or maps of the pre-Christian era have come to light, that could give some information on the development of cartography of that region in general and their connection with Indian cartography in particular. Their early maps were said to have been prepared with the help of palm leaves which is not considered a very durable material and, therefore, these might have decayed with time.[38]

Indian knowledge of geographical, astronomical and cosmological truths has its origin, as already stated, many centuries before in the pre-Christian era. For the reasons, given earlier, many classical works could be written only in the early Christian period. These would be touched upon in the concluding chapter.

Map 5: *Outline map of the Orbis Terrarum of the Romans* by Marcus Vipasanius Agrippa (12 BC). (Courtesy: Dr. R.P. Misra, *Fundamentals of Cartography*, Mysore, 1969).

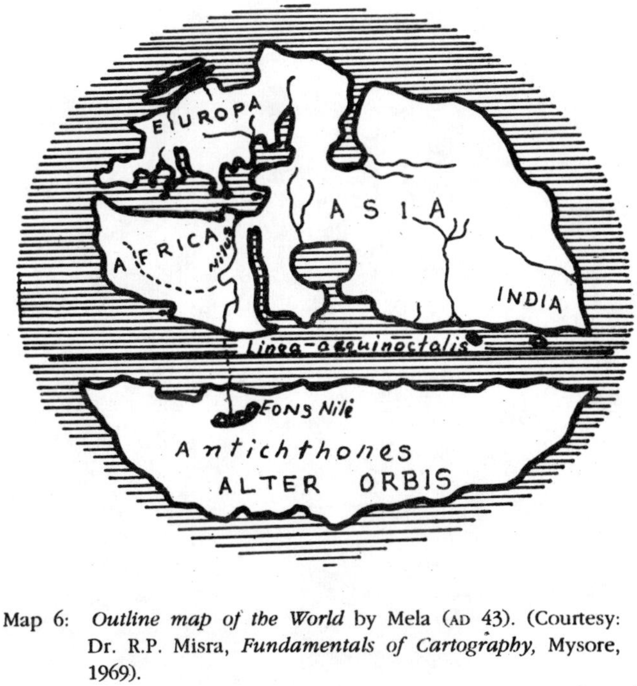

Map 6: *Outline map of the World* by Mela (AD 43). (Courtesy: Dr. R.P. Misra, *Fundamentals of Cartography*, Mysore, 1969).

NOTES

1. Leo Bagrow, op. cit., p. 31.
2. R.V. Tooley, *Map and Map-Makers*, London, 1970, 4th edn., p. 3.
3. Leo Bagrow, op. cit., p. 31.
4. Ibid., p. 31.
5. R.V. Tooley, op. cit., p. 3.
6. Leo Bagrow, op. cit., p. 32.
7. A. Cunningham, *Geography of Ancient India,* London 1871, p. xxi.
8. Leo Bagrow, op. cit., p. 33
9. S.M. Sastri, op. cit., p. xv.
10. D.F. Lach, op. cit., Vol. I, p. 5.
11. Ibid., p. 6.
12. S.M. Sastri, op. cit., p. xv.
13. Leo Bagrow, op. cit., p. 33.
14. D.F. Lach, op. cit., Vol. I, p. 6.
15. S.M. Sastri, op. cit., p. xv.
16. J.W. McCrindle, *Ancient India as Described in Classical Literature*, Washington, 1901, p. 1.
17. S.M. Sastri, op. cit., p. xv.
18. D.F. Lach, op. cit., Vol. I, p. 7.
19. Ibid., p. 8.
20. J.W. McCrindle, op. cit., p. xiv.
21. Narain Singh Kalota, *India as Described by Megasthenes*, Delhi, 1978, p. 19.
22. Ibid., p. 20.
23. D.F. Lach, op. cit., Vol. I, p. 10.
24. Leo Bagrow, op. cit., p. 33.
25. S.M. Sastri, op. cit., p. xvi. Quoted from E. Bunbury's *History of Ancient Geography,* Vol. I, London, 1883, p. 635.
26. D.F. Lach, op. cit., Vol. I, p. 13. Some others believe it to be 62 BC to AD 19.
27. Ibid., Vol. I, p. 11.
28. R.P. Misra, *Fundamentals of Cartography*, Mysore, 1969, p. 25.
29. S.M. Sastri, op. cit., p. xvii.
30. D.F. Lach, op. cit., Vol. I, pp. 13-14.
31. S.M. Sastri, op. cit., p. vii.
32. R.P. Misra, op. cit., p. 26.
33. D.F. Lach, op. cit., Vol. I, p. 14.
34. Leo Bagrow, op. cit., p. 197.
35. Ibid., pp. 198-9.
36. '*Xinhua*'—Chinese Daily newspaper, quoted in the *Indian Express*, Delhi edn., dated May 18, 1989.
37. Leo Bagrow, op. cit., p. 200.
38. Ibid., p. 208.

3. The Age of Reconnaissance
AD 1 TO 1000

With the change from pre-Christian era to the Christian period, a marked change in the cartographic scenario of the world is also discernible.

Several geographical discoveries and cartographic innovations took place during the first millennium of the Christian era.

During the closing years of the earlier era and the first century of the new one a brisk trade was springing up between the great mart at Alexandria and the western coast of India, especially the Malabar region. Alexandria had replaced the old Phoenician ports of Tyre and Sidon as the clearing house for Oriental goods.

Some of the eminent writers of the time were also ushered in to the new era. They wrote invaluable memoirs and treatises some of which are, unfortunately, totally lost. The most striking of them all was Strabo (63 BC-AD 21), although Diodorus and Plutarch also made their contributions.

The early writers of this era, like Pliny (AD 27-79), Marinos (*c.* AD 120), Arrian (AD 200), though none of them had slightest personal knowledge of India, benefited much from the writers of the pre-Christian era. Their works, though invaluable to history, are not of much real value for geography. Few of the places mentioned in them could be identified with any real approach to certainty.[1] Ever since Dr. Fleet expressed the above feelings, much water has flown down the Ganges, and modern scholars have been able to identify many of them.

As for Strabo's account of India, he himself had admitted that it could not be absolutely correct. In excuse thereof he had pointed out the difficulty of getting correct information about India owing to its great distance, and the fact that by then only part of it has been visited by a few 'ignorant' men who were not qualified to write an account of it.[2]

Pliny was known as a naturalist. He dealt with the geography of India in the sixth book of his *Natural History*. He referred to the increase in trade between Europe and the East. It was he who mentioned that the discovery of the periodic nature of the monsoons of the Indian Ocean was made at that time by Hippalus, which enabled the European navigators to take a direct route to India. Modern writers however dispute this 'discovery' by him. About Pliny's adherence to hard facts, it is said that 'hearsay-tales' of the sailors were given much credence by him whose 'love of the marvellous disposed him to accept far too readily even the most absurd fiction.'[3]

The tremendous increase of trade between Europe and India might have fuelled the need for some sort of practical merchant handbook called the *Periplus of the Erythrean Sea*. The book gives, in some detail, the Indo-Roman trade by sea at that time. The author of the *Periplus* had visited India and travelled widely. He stayed at Barigaza (Broach), which was the maritime terminus of east-west trade in India at that time for quite a long time. He described the entire coastal region of India, especially the port towns from near the mouth of the Indus river in the north-west coast covering the entire coastal belt up to the delta of the Ganges in the east. He also listed goods which the Romans got from the north-western Indian coast. It seems that the author had not visited Sri Lanka but had only heard of it from other sources. Again, the *Periplus* also gives details of trade with ports on the eastern coast of India, but his knowledge was from second hand sources only.[4]

The date of *Periplus* is much disputed. Rawlison states that the existence of monsoon winds, with the help of which vessels from Aden could sail directly across the Indian Ocean to the western Indian coast, was discovered by about AD 50. This is mentioned according to him, in the *Periplus of the Erythrean Sea*, written by an Alexandrian sea captain about the time of Nero (AD 54-68).[5]

But according to George Faldo Hourani, recent researches place *Periplus of the Erythrean Sea* in the later part of Ptolemaic period (323-30 BC), i.e. in the closing years of the pre-Christian era. He later becomes a little more specific and asserts that the discovery could not be later than 90 BC in any case. He further states that Hippalus did not discover the existence and behaviour of the monsoon winds, but only the way as to how this behaviour of the monsoon winds could be made use of by westerners in sailing across the ocean direct to the west coast of India; and that Hippalus is not the name of a person but was given to that portion of the monsoon winds that flowed from the Arabian Sea to the west coast of India during the summer season.[6]

Another recent, well-known and serious Belgian

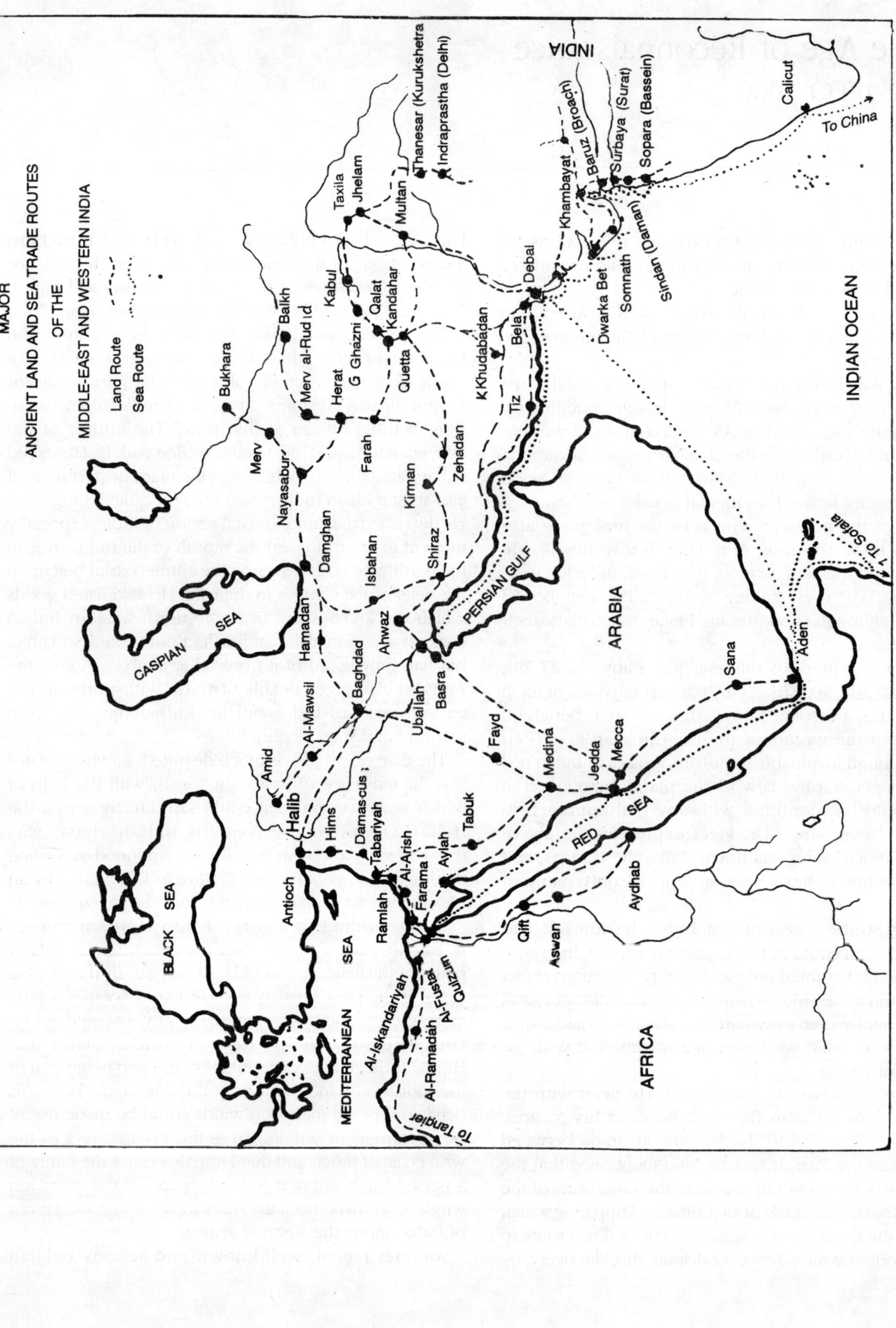

Map 7: *Major Ancient Land and Sea Trade Routes of Asia-Minor leading into India.* (Compiled by the author from various sources).
Note. Indian coast-line has been photographically reduced from the Survey of India 1:21 million Plain Outline Map of India, printed 1987.

scholar[7] of maritime history, trade and navigation of the early Portuguese, the Spaniards and the French, is of the opinion that *Periplus* which is the work of an Alexandrian merchant and which was written towards the beginning of the third century AD, has been very often misdated to the first century AD.

The question under discussion is not who discovered the behaviour of the monsoon in general but rather its discovery by the Westerners for the first time during the first century AD, or in other words, the Westerners came to know the mode of direct sailing during the first century AD, as this behaviour of the winds is stated for the first time by Pliny in his *Natural History*. Much of the confusion has been set right by him in his book *Natural History* (VI, Section 26, pages 100-1). The study of the relevant passage in the volume, which has often been misdated to the first century AD,[8] reveals the various stages of the so-called discovery of the western monsoon route to the east which could be stated thus: the first route from the promontory, Syagrus, goes to Patali at the mouth of the river Indus—almost a coasting voyage. Then, the sailors attempted from the same promontory towards the middle of India, i.e. the port of Zigerus, and finally, the third stage was from the south of the Red Sea, near the strait of the Bab-el-Mandel, from the port of Ocelis towards the Muziris or the south India. All these three stages were known to the Westerners during the reign of Claudius (AD 41-54) as an embassy from Sri Lanka (Ceylon) was sent to his Court. Further, prior to this, commercial navigation or maritime contact was at its peak between Rome and India.[9]

A detailed study of the goods imported from India and the regions from where these were being exported as given in the *Periplus*, along with the years in which this was being done together with the fluctuations in the quantum of the flow of goods, along with the description and the use of different types of vessels or ships on various routes and, finally, its comparison with the recent researches on the maritime history and commerce seem to indicate that the *Periplus* could have been written not in the period to which it has generally been ascribed to by early writers, i.e. first century AD, but somewhere in the third century AD.

During the early second century AD, there appeared a writer whose work had almost an undisputed authority in the field of geographical knowledge of the world and map-making for more than a millennium. His name was Claudius Ptolemaeus, generally famous as Ptolemy. Virtually nothing is known about his early life. According to Theodorus Meliteniota of Byzantine, Ptolemy was a Hellenized Egyptian and was born in the Grecian city of Ptolemais Hermii. The year of his birth is not known but the period during which he was active can be ascertained from the dates of certain astronomical observations which he made and which were recorded by him in his celebrated astronomical work *Almagest*; this period on a rough estimate spans between AD 127 and 160. Arabian traditions make us believe that he died at the age of 78. We may, therefore, safely say that he flourished and worked during the second century AD.

Ptolemy was a celebrated astronomer, mathematician and geographer. His reputation as a geographer was mainly due to his *Guide to Geography*, or *Geographia* or even *Cosmographia*, a work of diligent compilation and not a work of original thinking,[10] analysis and deductions. His scientific approach to geography exerted a great influence on future generations. It is not a descriptive work either but an exhaustive gazetteer of places arranged regionwise with latitude and longitude assigned to each place. But surprisingly the *Geographia* made its impact upon the Westerners much later, nearly about 1000 years after it was written.

Ptolemy derived his knowledge about India and Asia from the earlier works of Greek writers and from the oral reports of traders and navigators who used to visit his native port city of Alexandria[11] which was already well-known as a centre of east-west trade and geographical speculations. He is not explicit in saying whether he made any map himself. It appears that his main concern was to lay down instructions for map-making, study of projections and making of regional maps from a world map and detailed listing or cataloguing of latitudes and longitudes of fixed points or places that would serve as a base for a map.[12] By the closing years of the Byzantine Empire a number of manuscript copies of Ptolemy's *Geographia* had come out, some containing maps and others without them. Even those that contained maps, differ from one another. In some manuscripts there were 26 maps and in some others 64 maps. The scholars, even now, have divergent views whether the originals of these maps, since these were claimed to be copies, were made by Ptolemy himself or by someone else, or whether the copies were made during Ptolemy's lifetime or were prepared much later.

It is now agreed that the Byzantine manuscript copy was the earliest copy known. It consisted of eight volumes or Books. These volumes contained criticism of some cartographers regarding their methods of construction [of maps, detailed lists of various coordinates] and other data and instructions for using the material for map-making, etc.[13]

Doubts about the originality of the maps that accompanied various copies of the *Geographia* are due to inconsistencies and frequent contradictions in the various copies of the texts, due to inclusion of material that could not have been known in Ptolemy's time, and also the fact that the maps do not collate with the texts and thus seem to be of later period. According to Leo Bagrow, Ptolemy himself had perhaps written, Book I. Some parts of Book II, the information regarding the compilation of regional maps only and possibly, chapters 3-28 could also be ascribed to him. Slowly and gradually, during many years after the *Geographia* was written by Ptolemy, later geographers added little by little to it before the work took its present form and finally, was put together in eight volumes or books.[14]

A number of new copies, manuscript as well as printed, came out in subsequent years with each writer or publisher either incorporating 'the latest information' to the text and maps or putting altogether new 'tabula' along with the original text with maps based on new information.

Leo Bagrow is of the opinion that the *Geographia*, in the present form, was evolved in three or four stages and was ultimately put together during the course of twelve centuries. It was written in the second century and was put together by the fourteenth century.

In spite of the fact that the *Geographia* in the present form was not his handiwork alone but the result of cumulative experience of many geographers, Ptolemy's fame has not diminished. This was because he was known also for another famous work—work in astronomy—known as *Syntaxis* which was well known in the Arab world as *Almagest*.

Significantly such an important geographical work was not known to any of his contemporaries, nor any writer of the following few centuries; even the dictionary of ancient authors that was compiled in the tenth century by 'Suidas' was silent about it.[15]

Ptolemy was a compiler. He got his ideas about the measurement of the earth from Marinos of Tyre during the second century AD, who, in turn, got them from Posidonius. It may be remembered that Hipparchus had propounded the theory of dividing the earth into degrees of angular distances and also using of coordinates about three centuries earlier and Ptolemy formed his theory of plotting a map from all this. But at the time when Hipparchus made his suggestions, little information about the coordinates of places was available and, perhaps, because of this, he seemed to have used, as an alternative, the method of dividing the then known world into zones or *Klimates*. Ptolemy, therefore, started compiling his catalogue of coordinates, of places, etc., from various *Periploi* available to him for the Mediterranean sea, from Marinos and also, perhaps, from Strabo's works, though Ptolemy had given very few descriptive details in his *Geographia*.

As a compiler, Ptolemy had to select from various forms of data available from many of the early writers on the subject. Unfortunately he unwittingly selected Marinos' data and method for the measurement of the earth which ultimately caused unaccounted distortion in his calculation of the size and shape of the earth.[16]

It is curious that such an important a work of geography and cartography, written in the second century AD remained almost hidden from the later geographers for about six centuries, when ultimately, it was translated into Arabic in the eighth century. Soon, additions were made to it by Al Kharizmi.[17] For Europeans, it came out from oblivion so to say after more than a millennium after its creation, when it was translated in 1409 into Latin.[18] Subsequently, numerous editions were brought out in various languages in Europe in the following two centuries. The scholars of the early fifteenth century Europe were dazed at what Ptolemy had written as it was something quite new and exciting. They were in the same kind of mental situation as they had been when they came to know of the exploits of Marco Polo and the things he described or narrated.[19]

Out of the twenty-six regional maps, twelve maps are on Asia, India is shown on the tenth map, known as 'Tabula Asiae X' of Ptolemy. The coastline of India, as shown in this map, is more or less in a straight horizontal line with no peninsular formation at all and Sri Lanka is shown, in proportional comparison to India, as a huge island in spite of the fact that many people earlier when Ptolemy wrote his *Geographia*, had visited India and had described its shape. All this information must have been available to him in the world's renowned library of Alexandria, and the travellers and navigators who frequented the city must have described India and India's coastal regions if he had cared to enquire from them. This glaring discrepancy did not come to his notice perhaps because the maps, the later incorporations with the text, were not prepared by him. Also possibly by going through the lists of the coordinates, latitudes and longitudes, of the place names on the coastal India that he compiled and listed, he could not visualise the shape of India as a whole or did not give any heed to this aspect. This was more so because of the basic miscalculation on his part that he trusted Marinos' idea about the measurement of a degree which contributed largely to his inability to imagine the correct peninsular shape of India without having maps in front of him. Or else in all probability, he would surely have revised his whole concept.

In spite of these and other shortcomings, the *Geographia* dominated the geographical thought of Europe and Asia Minor for more than 1,500 years. The method which he adopted for measuring, dividing and mapping the world, surely provided the later geographers with, more or less, scientific base for future advance in the field.

There is another category of 'writers' more prominent in this era who did not 'write' about geographical description of a country, but used only sign sequences or words and had drawn bare outline of coastal areas, even sometimes in a crude form. Their works were called sea charts. As already seen earlier, ancient navigators followed the method which may be called close-to-the-shores, or, in other words, the land was never allowed by them to be out of sight. Then came the period when the behaviour of oceanic winds was mastered and used for open sea navigation. Gradually, with the development of navigational aids or evolution of methods of finding one's bearings in an open sea, far away from the land, navigators needed oral or written instructions of some sort to guide their vessels and also had to remember the outlines of landforms to recognise the coasts. Earlier guides to navigation, wherein no maps or charts showing landforms were included but contained only written sailing directions from one particular point to another, were called *periplus* by Greek writers or 'coast-pilots' by sailors generally. The oldest extant *periplus* is said to be by Hanno, of 450 BC, but the oldest sea chart is of about AD 1300. A number of such guides to sea and land routes have come down to us from the Byzantine era,[20] which are the oldest ones of the kind. Some itineraries of seafarers containing geographical descriptions of the land they travelled, also contain sailing directions or details to such a great extent that these could serve as guides to pilotage. For instance the itinerary of Nearchus from the mouth of the Indus to the Shatt-al-Arab, describing the course of the ship in great detail, gave almost precise distances from one stage to another, as in a *periplus*, including wind directions and its vagaries in a particular season and details of the coastlines, sandbanks and even dangers from under-water rocks, other navigational hazards near the coasts, etc. Another form of sailing directions is recorded in Pliny's *Natural History*, Book V where instead of giving wind directions, native name of a particular wind at that particular point is mentioned. As the sailors were used to these names and were familiar with their behaviour in a particular month, this served as guide to the sailors in that region at that particular time of the year. With the invention of compass, old *periplus* gradually gave rise to another type of guide or sea chart known as *portolano* or portolan. These developed further through the tenth century to early sixteenth century AD. In the later period, these included rhumb-lines or loxodromies network and compassroses along with a little more details of coasts.

After the dissolution of the Roman Empire, and the consequent fragmentation of it into small units by the middle of the seventh century AD, Arabia became a great power. Once it was established, it led to the revival of interest in geographical, cartographical and astronomical pursuits. The conquest of Syria, Mesopotamia, Egypt and other kingdoms in the region brought the Arabs into direct contact with superior civilizations of the Asia Minor and the East. For about six centuries hence cartographical and geographical knowledge of the rest of the known world came to Europe mainly via the Arab centres of learning—Baghdad and Damascus. The astronomical, cartographical and geographical speculations of the intellectual centres of Europe in the thirteenth century owe a great deal to Arab advancement. Thus it was the Arabic cartography that worked as a link between the ancient learning of the antiquity and the European cartographic revivalism.

The religious obligations forced the Arabs to study mathematics, astronomical geography and cartography to enable them to calculate accurately the latitudes and longitudes of places. This was required to ascertain the direction of their sacred city of Mecca. Their calculations of geographical coordinates were much more accurate than those calculated by Ptolemy.

Arabs in general adopted in the early stages, Ptolemaic ideas of landlocked terrestrial world in the south and limitless ocean (Atlantic Ocean) in the west. For various reasons, both political and commercial, the Arabian scholars postulated this theory.

During the eighth century Caliph Al-Mansur (AD 754-75) ordered Mohammad ibn Ibrahim, known as al-Gazari, to compose astronomical tables (AD 772-3), to translate and re-cast Indian tables and other astronomical works. He also introduced the Arabs to astronomy. Being a learned scholar himself, he composed *Sind-Hind*, during the ninth century on the basis of Sanskrit *Siddhantas*. This work became very popular.[21]

It was during the reign of Al-Mamun (AD 813-33) that manuscripts and treatises available in all the advanced languages like Persian, Armaic, Sanskrit and Greek were translated into Arabic. Arab geographical knowledge is said to have expanded after the Arabs had established themselves in Baghdad. It was the shifting of their capital from Damascus to Baghdad that brought them in very close contact with the merchants and scholars of the East. With the expansion of the Arab empire their geographical knowledge also widened.

Arabs were great travellers as well. Early in the eighth century, route maps and itineraries were compiled mostly for war purposes and the movement of diplomatic missions to various countries, as far as China, etc.

Many of the travel accounts of the Arabs were published in the ninth century. Khwarizmi in his work *Face of the Earth*, also improved upon Ptolemy's *Geographia*. Sulaiman's many voyages to India also took place during this period. Ayub Yusuf Yakub authored more than 200 books which also include translations of Sanskrit, Persian and Greek works along with that of Ptolemy's *Geographia*. Yahya published many astronomical treatises and charts including tables of coordinates, etc. Abu Mashar (AD 805-85), a versatile scholar, studied almost all the astronomical works of the pre-Arab-Persian period, and prepared his own astronomical tables. He had mastered the behaviour of the winds, tides and sea currents of the Indian Ocean and other seas frequented by Arab navigators and traders during the ninth century. Knowledge gathered by eye-witnesses forced the Arabs to break away from their traditional beliefs, revise existing theories and subject their conclusions to severe analysis and criticism. The study of ancient works in Persian, Sanskrit and other languages gave them the necessary guidance and knowledge to negotiate the open seas and enabled them to describe positions of the places accurately.[22]

As already stated, Al-Mamun was himself a great scholar and it was during his time that Ptolemy's map was improved upon. Muhammad ibn Kathir Al-Fargani, the Arab scholar, who died in AD 830, made Ptolemy's *Geographia* more popular. He was also responsible for the introduction in the Arabic cartography, the division of the inhabited world into climatic zones. Two well-known works of Ptolemy were translated into Arabic—*Syntaxis* was called *Almagest* and *Geographia* was entitled *Kitab Gagrafya*.[23]

By the order of Caliph Al-Mamun, Muhammad ibn Musa Al-Khwarizmi (*c*. AD 820) wrote a descriptive note to a world map, also probably prepared by him on the basis of Ptolemy's *Geographia*.[24] Apart from some more individual maps in the ninth century, two more translations of Ptolemy's *Geographia* were made but unfortunately both are now lost. During this period, a Syrian scholar, in his work *The Six Days*, incorporated a T-O world map.[25] The map is south-oriented with Mount Sinai in the centre. There is another similar map in the Arabic translation of St. Isidore's *Etymologiae* of the tenth century.

Yet another important work based on Indian writings, mostly on Mathematical-geography and astronomy came out sometime during the period between the end of ninth century and the early years of the tenth century. Its author was Al-Bhattani.

During the tenth century, there were five prominent authors—Abu Zaid Al-Balkhi, who was the first to write during AD 919-21; Al-Istakhri, wrote in AD 934; Al-Masudi, who died in AD 956; Ibn Haukal, wrote in *c*. AD 980, and Al-Muqaddasi in AD 985.

Al-Balkhi wrote *Figures of Climates* wherein he described climates of all the regions of the known world. He also incorporated maps, one map covering one region. Al-Istakhri travelled far and wide in AD 951 from Atlantic to India. His work, *Book of Routes*, served as the basis for many future writers to improve upon. Ibn Haukal did this in AD 977 in his work *Book of Routes and Realms*, though both depended much on what Al-Balkhi had written earlier.

Out of these five writers of the tenth century, Al-Masudi is one of the best known geographer of his times. He made many journeys to India and other countries. He wrote widely without fear or prejudice and even wrote against many of the prevailing ideas propagated by Ptolemy's *Geographia*. He firmly wrote that Africa was surrounded by a sea. This idea was against the prevailing beliefs. His most important work was the *Annals of Time*. Unfortunately, this work is not extant now. In his other work, called *Meadows of Gold*, he compared various theories regarding the origin of the Earth propounded in Greek, Indian and Sabian countries. Al-Muqaddasi authored *Best Classification for the Knowledge of Climates* during the tenth century.[26]

Except Al-Masudi, all the other four scholars included in their works the modified and improved versions of the T-O maps. Their maps looked like illustrated itineraries of route maps. The collections of maps of these four scholars look almost similar in appearance and content. For example, each contained equal number of maps of the same regions and areas and equal number of sea charts as well, along with of course a standard type of almost similar text. These collections of maps were generally called 'Atlas of Islam'. From what has been noted above about these atlases, it could be concluded that these originated from one source or writer and others just followed or copied them with slight changes made every time these were copied by the others. Ibn Khurdadbah was, perhaps, the first author of such an atlas, but this did not contain maps. It was prepared in the ninth century. These atlases underwent more changes in the second millennium AD.[27]

The rise of the church power in Europe created a peculiar situation. It crushed the secular culture and ideas of Greece and Rome which, after some years, were almost forgotten. The church had not developed its own views about cosmology. This attitude affected the cartographic development[28] adversely. The scholars were not ignorant, however, about secular cartographic heritage as their oversimplified maps would make us believe. Because of the religious compulsions of the Church, their aim might have undergone a change and were then catering to the ignorant masses; their maps were more symbolic rather than utilitarian. Their *mappae mundi* or world maps were, as their traditions would have it, only used for illustrations in historical or descriptive works, or, we may say, the maps were used for illustrating their religious dogmas.[29] They had given a peculiar twist to the mapping of the known world. Their concept of the world was diagrammatic, the three continents known in the ancient world being symbolised by the placing of a 'T' within an 'O'. These maps were mostly east-oriented. Occasionally, in some early maps, we find the same idea represented in a somewhat different form—a triangle or a square—with a 'Y' written within an 'O', or a 'V' within a square. Generally, the world was depicted as a circular disc, with Asia allotted the upper half, and Europe and Africa, the two lower quadrants. The horizontal line separating Asia from Europe and Africa, representing the horizontal bar of 'T', generally passed through the Don (Tanais), the Aegean to the Nile onwards. The line passing vertically through the Mediterranean Sea meeting the horizontal line, formed the vertical stroke of the letter 'T'. Such maps were known as T-O maps. As already stated, these maps were very common during those days.[30]

Another type of map available during this period was called Crates' map. This map was rare because the Church did not agree with the basic theory it represented. Such maps more mathematical in conception, accepted the sphericity of the world, and portrayed a single hemisphere, divided into five zones or seven climates of the habitable world, the whole encircled by an ocean. These maps were generally known as the hemisphere or zone maps. Martianus Capella and Bishop Theodulf made use of such maps to illustrate their works.[31]

Later in the twelfth and thirteenth centuries and again in the fifteenth century, zone theory was further developed, mostly by Arabs.

There is yet another type, a mixture of the two types mentioned above. This was developed in the early second millennium.

In the first millennium of the Christian era, the Chinese cartography made considerable progress, especially after the invention of paper in AD 105. The new maps, though mostly representing Chinese regions, were administrative in nature.

As already stated earlier, Buddhism influenced Chinese thought, cosmogony and cartography, but this influence is visible only in a few Chinese maps, chiefly those that are incorporated in texts based mostly on Indian thoughts.[32] In AD 793, Chia Tan contributed much by preparing regional maps and itineraries.[33]

Chinese contribution to Indian cartography, directly or indirectly during the first 1000 years of the Christian era is more or less non-existent.

NOTES

1. Dr. Fleet, *Indian Antiquity*, 1901, p. 284.
2. J.W. McCrindle, op. cit., p. 9.
3. Ibid., p. 102.
4. D.F. Lach, op. cit., Vol. I, pp. 14-15.
5. H.G. Rawlison, op. cit., p. 245.
6. G.F. Hourani, op. cit., p. 24.
7. Charles Verlinden, was Hon. Chairman, International Commission for Maritime History, Brussels, Belgium.
8. Charles Verlinden, op. cit., p. 32.
9. Ibid., pp. 32-4.
10. J.H. Parry, *The Age of Reconnaissance*, London, 1963, p. 10.
11. D.F. Lach, op. cit., Vol. I, p. 16.
12. Leo Bagrow, op. cit., p. 34.
13. Ibid., pp. 34-5.
14. Ibid., p. 35.
15. Ibid., p. 35.
16. J.H. Parry, op. cit., p. 12.
17. R.V. Tooley, op. cit., p. 9.
18. *Encyclopedia Britannica*, 1967, Vol. 18, p. 813.
19. J.H. Parry, op. cit., p. 13.
20. Leo Bagrow, op. cit., pp. 61-2.
21. Ibid., p. 53.
22. V.A. Janaki, op. cit., p. 8.
23. Leo Bagrow, op. cit., p. 54.
24. Ibid., p. 54.
25. T.O. maps are described later in the chapter.
26. V.A. Janaki, op. cit., pp. 5-6.
27. Leo Bagrow, op. cit., p. 55.
28. Ibid., p. 41.
29. R.V. Tooley, op. cit., p. 12.
30. Leo Bagrow, op. cit., p. 43.
31. R.V. Tooley, op. cit., p. 12.
32. Leo Bagrow, op. cit., p. 198.
33. Ibid., p. 199.

4. The Age of Adventure and Discovery
1000 TO 1600

The rise of the Arab power since the seventh century, and invasions from the Mongols and the Turks in the Middle and Near East created unsettled conditions. Consequently the traditional sea route to the East was blocked. Further, destruction and dislocation due to Hundred Years' War and the epidemic of the 'Black Death' in Europe gradually dislocated the economy of the European nations. This gave spurt to the price rise of spices, silks, muslins, etc., that were in great demand. The trading nations tried to cash in on the situation and made hectic efforts to find new sea routes to the East, as the land routes were closed to them. While the Portuguese looked for a route around Africa, the Italians tried to break the stranglehold of Egypt and the Turks through diplomacy and even war.[1]

Much before the time when the necessity of finding alternate route to India arose, during the twelfth century a great Arab scholar of Mathematics, Al Fragan was calculating the length of a degree, an intricate problem in cartography in those days. His figure came to $56\frac{2}{3}$ miles, whereas Ptolemy's figure was $62\frac{1}{2}$ miles. With these figures the circumference of the earth was calculated to 24,480 miles and 22,500 miles for Al Fragan and Ptolemy respectively. Thus Al Fragan's calculation was much nearer the true one. D.C. Sircar has given the details in a very lucid manner in his illuminating work.[2] It is this mistake made by Ptolemy, about the length of the degree, that resulted in the distortion of his map.

As stated earlier, the simplified map of the world was further simplified and modified by Al Biruni in AD 1021. In his work, Al Biruni described the various seas like the Western Sea (now Atlantic Ocean, he called it as Environing Sea), spreading to the north and the south of the globe. According to him, Africa on the western part of the Southern Sea, now known as Indian Ocean, had been reached by the eleventh century AD, up to Sofala, on its eastern coast, about 20° south latitude. Again, he asserted that Africa was surrounded by the Environing Sea.

However, it was Idrisi who showed remarkable perception of the world. He had travelled extensively over Europe, Africa, Levant and the East and finally settled down in Sicily. Born at Ceuta, now Morocco, in 1100, he worked at the Court of King Roger II (AD 1097-1154) who had invited him. Roger himself was a great scholar. He asked Idrisi to compile geographical data, such as coordinates of every place he had visited on his travels, different climates, various gulfs in the region, etc. Idrisi even sent some persons to collect whatever information he missed to complete the book. He afterwards compiled a map of the world also. This book was later translated in Latin as well. The Arabic version and its Latin translation, along with the map, were completed in 1154. Idrisi named the book after his master *Kitab Rudjar*, the Book of Roger, and the map, as 'Tabula Rogeriana'. Roger, on his part, called the book as *Nuzhat al-Mushtak*.[3] Idrisi's work remained an authority on the subject for about four centuries. As late as the mid-sixteenth century, Sfax of Tunisia, a famous family of cartographers called Sharfi, produced maps which were mostly based on Idrisi's. His maps could be said to be the last bastions of the Arabic cartography.

In the fourteenth century Catalonia rose as a Mediterranean commercial power which posed a threat to Venice and Genoa in their trade with the north African nations.[4]

Genoese, Catalans and Portuguese, all tried to sail through the Pillars of Hercules (Gibraltar), down the African west coast, in search of an alternate route by going around it, but failed. It was ultimately left to Bartholomeu Dias, a Portuguese, to successfully negotiate around the Cape of Good Hope in 1487. This was the result of systematic and continuous explorations of the African west coast under the guidance of Prince Henry, the Navigator (AD 1394-1460), in the early fifteenth century.[5]

As the Portuguese were busy in their efforts to master the sea route to India around Africa, other Europeans occasionally made their way to India along the old routes. Some made their way through circuitous land routes and came to India the other way round, like the now famous, Marco Polo. They all have left vivid geographical accounts of India which are now extant.

Land routes from Europe to Mongolia, were opened in the mid-thirteenth century. This was followed by the extension of Mongolian hegemony over China in 1264. Kublai Khan, the Mongol ruler, took up his residence at Cambaluc (Peking). He received, among his first visitors from Latin Europe, the Polo brothers, Nicolo and Maffeo. There were Venetian merchants who followed the trade routes to Cambaluc. After some time, these two Polos were sent back by Kublai Khan, as his emissaries to Europe with a request to the Pope to send some learned men to Cambaluc to instruct the Tartars about European way of life and culture. Pope, in return, appointed the Polos as Apostolic delegates to the Khan's court. They set out for Cambaluc in 1271. On this journey they were accompanied by Nicolo's younger son, Marco. They reached Cambaluc in 1275.[6] Young Marco Polo, in all probability, if his accounts are to be believed, was, in some official capacity, able to travel from one end of China to the other. During his travels he used to take notes for the Khan who was mostly interested in topography and the people. He came back to Venice in 1295. His book, *The Description of the World*, was written in 1298-9. Marco dictated his story to Rustichello of Pisa, a writer of popular romances. The book was widely circulated, translated, and adapted, even before the author's death. There are 119 manuscript versions extant of *The Description of the World* in various languages, no two of which are exactly alike. This was the reason, perhaps why many of his contemporaries suspected the accuracy of Marco's account. On the other hand, it is stated that on his deathbed he told the persons present there that he could not tell all that he knew about China and of other places he had travelled. He knew so much that he had to select what to say and what not to tell, thinking it, perhaps, not worth it. It was this selective nature of events he disclosed, that was responsible for a number of omissions that the later writers noticed in his accounts.[7]

Looking back, it can safely be said that Marco Polo was an accurate and factual writer. Of the things and incidents, which he did not see or witness himself, he used his judgement very judiciously and he has been proved accurate.[8] In 1292, Polo travelled by sea to the western coast of India, Malabar. His few remarks about Indian navigation have already been commented upon earlier. It is curious that Polo's contributions to the geography of Asia had not been incorporated into any contemporary map of western origin until the construction of the Catalan map of 1375. But it is still more curious to know that the Chinese maps showed the places he visited in India in detail.

In the thirteenth century, Roger Bacon was the only writer who believed in the scientific analysis of literary evidence. He had deeply studied the works of Arab writers and their translations of even the ancient classical works in almost all the advanced languages of the world. He firmly believed that Asia and Africa extended south of the equator and also that torrid zone was habitable as well, which was contrary to the ideas held by Al-Masudi and others. The geographical sections of his classical work, *Opus Majus*, which came out in 1264, greatly influenced the thoughts of Cardinal Pierre d'Ailly, one of the greatest scholars of his times. The *Imago Mundi* of d'Ailly, was published in the late fifteenth century, in 1483, more than two hundred years after Bacon's book had hit the scene.[9]

In the mid-fourteenth century, a Spanish Franciscan wrote a book *Book of Knowledge of All the Kingdoms, Lands and Lordships in the World*. Whether the Franciscan had actually travelled to India, is disputed, but his knowledge of geography, his accurate naming of the cities, and his information about the land and sea routes, now seems to be as accurate as it could be in those days.[10]

During the next century appeared an account of another famous traveller, who spent about twenty-five years in the East, Nicolo de Conti, a Venetian. On his return to Italy in 1441, he was very closely questioned about his Eastern travels by the Papal Secretary and Humanist, Poggio Bracciolini, in Rome. Conti's travels took him to widely scattered places in India also. With such a long experience in the Asiatic region and voyages in the Indian Ocean and consequently his familiarity with Indian navigation, Conti was able to give intelligent comments on geographical and navigational matters. Without paying any attention to earlier topographical descriptions of India by others, he divided India into three major regions based on the natural divisions created by the mighty rivers, Indus on one side and the Ganges on the other. He was, perhaps, incorrect when he remarked that Arabs and Indian navigators were then not using compass as they did not know of it, but were navigating generally by looking at the stars visible in the southern part of the equator.[113] Although Conti's contribution to India's geography was not so extensive, but his few comments were forthright—his views on Indian culture, trade and commerce were quite detailed. Conti was the first European to travel extensively into the interior of the Indian peninsula, and to identify and describe number of cities in the Gangetic Valley as well. This reminds us of almost similar description by Megasthenes earlier in the last century of the pre-Christian millennium.

Bracciolini Poggio, a famous writer of the fifteenth century, gathered material for his great work *Historia de varietate fortunae* from 1431 to 1447, which came out in four volumes, volume IV of which concerns India. This book was not based exclusively on the information supplied by Conti, though his contribution was considerable so far as Asia was concerned. Conti's account was later published by many authors and under different titles, and was translated into many European languages. Many cartographers also used Conti's information on their maps of the Asian region. Without question, Conti's account had considerable influence on contemporary thought about geography.[12]

Yet another well known writer of the fifteenth century was Aeneas Sylvius Piccolomini. He was Pope Pius II during 1458-64 and was the Ptolemy's greatest critic of the day. His work—*Historia rerum ubique gestarum*—was published in 1461. He believed that India could be reached by sailing round Africa. As he was Pope at the time, his opinion carried considerable weight.

Again, another writer whose works had bearings on Indian geography, though indirectly, was Toscanelli, a friend of Poggio, who was one of the few geographers to repudiate in his writings, most of what Ptolemy and Strabo had written about Asia, in the light of later information available through Marco Polo, Nicolo de Conti and others. He sought to establish a realistic picture of Asia.

In 1487, King John of Portugal sent Pedro de Covilha and Afonso de Paiva on an inland trip across north Africa to seek out the trade routes to the East among other things. They reached Aden. From there, Covilha came to India by travelling ostensibly in an Arab vessel but de Paiva was never heard of again. Covilha sent a report through a messenger to King John II at Lisbon in 1490. His report contains details of trading centres in India, and his remarks on the mode of navigation[13] and also the trade route being frequented in the Indian Ocean, to which he was an eyewitness, were believed by the masses.

Of the last of the independent land travellers before the opening of the sea route was the Italian, Ludovico di Varthema. The beginning of his travel almost coincided with the opening of the sea route.

Varthema arrived at Cambay by the end of 1504, travelled extensively in southern India and came to Calicut at the beginning of 1505. He then proceeded southward, rounded Cape Comorin by sea and, perhaps, then travelled northward along the eastern coast of India. He visited Tennasserim, Pegu, Malacca, Sumatra, the Moluccas, and also Java. The information gathered on his sojourns to these regions was disputed by many writers. Like most of the other writers of the period, he paid more attention to Calicut and the Malabar Coast than to any other part of the East, and commented mostly on ship-building and navigation on the western coast. Varthema's account, mostly of the regions he visited after leaving Cape Comorin, came for severe criticism by other writers who followed him. The first scholar to challenge him was Garcia da Orta, the Portuguese naturalist, who argued that Varthema never went east of India and thus his account was from hearsay. This aspect of Varthema's information is still under dispute,[14] even though, according to others, his, on the whole very exciting and almost convincing, narrative of the journey was published in 1510, and was an immediate success.[15]

By this time, the Portuguese had become apprehensive of the competition from other European nations and wanted to keep to themselves the details of their overseas discoveries. They tried misinforming other nations by distorting the news deliberately about the newly discovered lands.

In 1504 King Manual, issued a decree that complete secrecy under pain of death, should be kept with regard to the voyages made through the southeastern and northeastern waters. On the promulgation of this decree, all the charts, maps, and log-books with regard to Africa, India, and Brazil were centralized and kept under strict control in the royal chartroom.[16] This secrecy was maintained so much so that not a single work was published during 1500 and the mid-century regarding their new discoveries. This policy of secrecy operated successfully during the first two decades of the century. Around mid-sixteenth century, the Portuguese control over their overseas empire began to crack. On the death of King Manual in 1521, certain amount of information about their foreign colonies, their sea routes and other secrets had begun to be passed on to the other nations by spies, foreign nationals serving on the Portuguese ships, merchants and also those Portuguese who went abroad to work there. The Jesuits who were there in the East by this time in quite good number earlier did not publish their letter books during the sixteenth century out of regard for the wishes of the Portuguese. However, during this century, nationals and governments of non-Iberian countries were also taking keen interest in these discoveries and were themselves trying to penetrate the Portuguese barriers with some success. In 1522, the successful circumnavigation of the world by Magellan's flottila, although only one ship could manage to complete the voyage round the world, proved to the Portuguese that they could no longer maintain the secrecy of the sea route. This news of the Spanish success ran like an electric current throughout the western Europe, and many nations began to cash on the news. The story was first published by Maximilian who interviewed Juan Sebastian del Cano, the Captain of the only ship 'Victoria', of the Magellan's flottila, who could manage to reach home. Another contemporary was Antonio Pigafetta, who accompanied Magellan as a chronicler of the voyage. He kept a running diary and wrote short accounts initially in 1522, but no record is now available, if it was ever published, although his manuscript copy was ready by 1524.[17]

By 1527, French vessels also appeared in the Indian Ocean and their first publication came out in 1531 in the form of poems; their journal was published as late as the nineteenth century in 1832. Thereafter, regular information about the East began to come out from French sources also. By then, the circle of interest in the overseas world widened in Europe.

The Dutch were not far behind; close to the end of the sixteenth century, the work of J.H. Van Linschoten, the first of the great Dutch writers, aroused the interest of the Dutch masses in the voyages of their countrymen to the East.

Once the barriers to the travels in the Indian Ocean were removed after 1550, materials on overseas world began to flood Europe. The earliest of the collection of travel literature was that of Giovanni Battista Ramusio (1485-1557), a Venetian, which was published in three volumes in 1550. The first volume of the work contained new information on Asia and was republished five times with additions, the last being dated 1613. This included among other topics, the earliest accounts of connections with the west, and activities on the Indian coast in 1537-38. It also contained Italian version of the *Periplus of the Indian Ocean* written by Flavius Arrian—along essay by him on the ancient and sixteenth century knowledge of the shores of the Indian Ocean, account by Barbosa (details of D. Barbosa's and Pires' accounts were state secret and were not allowed to be published in Portugal), etc. This volume included a map of India as well. To cosmographers, cartographers and historians interested in the events in the sixteenth century, Ramusio's collection has proved to be an essential source material.

Likewise, the pioneer collector in England was Richard Eden. He was greatly influenced by the exploits of other European nations in the East and wanted Englishmen as well to trade in the overseas world. To awaken the English masses to this end, he began to compile travel literature and translate the same from other languages. His first publication came out in 1553. It was titled *A treatyse of the newe India* and included, among other articles, a translation of *Cosmographia* by Sebastian Munster, which was published by him in 1540. This was a great success amongst the English scholars. Another of his books, *The Decades of the newe worlde or west India*, was also published in 1553. It mostly contained the exploits of the Spanish writers. The preface, written to this volume, at once became famous and is still considered to be a masterpiece of political propaganda.[18]

By this time Europe, especially England, was reeling with excitement over the exploits of Francis Drake who had gone round the world in 1579. People were eager to know the details of this historic voyage which followed the south-westerly route, rounding the strait of Magellan, almost at the southern tip of the South-American continent.

One of the most vivid and assiduous English collector of information on Asia, including India, was Richard Hakluyt. After graduating from Oxford in 1574, he developed interest in geography. He took a postgraduate degree in 1577 and immediately began giving concrete shape to his life's ambition. Very early in his career, he started interviewing people in search of information, collecting manuscripts and books on geography and cartography, on travel and navigation, and even on commercial problems, the information that was necessary to make the English nation interested in the East. In his efforts in this direction, he was greatly helped though indirectly, by the only English Jesuit residing in Goa, Thomas Stevens (he is also referred to as Thomas Stephens), who became Rector of Jesuit College in Goa. He wrote a detailed letter to his father in England. A copy of this letter was made available to Hakluyt. He even met the well-known cartographer, Ortelius, in 1577 and corresponded with Gerard Mercator as well. Richard Eden and Richard Hakluyt were the main force behind the motivation of the English nation to take interest in the overseas world trade. Ultimately, he was successful in forcing the trading classes to come together and form a company to trade in India in 1600.

To have an overview of the sixteenth century European source materials on India the following were noteworthy—the Portuguese secular writings, their chronicles (which were best on the trading ports on the western coast of India), the Jesuit letter-books, histories and the writings of the travellers from non-Iberian region like English, Dutch, German, French, etc. These varied writings complement as well as supplement each other. The story of the European exploits in the East till the sixteenth century, as given out by early navigators and travellers, was continued by the Portuguese chronicles during the Portuguese hegemony over the Indian Ocean. These chronicles came to an end around 1540. Further exploits were revealed to the people by the Jesuits. This source of information somewhat dried out in 1560s and then further story was taken up by independent merchant-writers and commentators like Cesare Fedrici,[19] Ralph Fitch[20] (1583) and others. Obviously, Jesuit letters were more or less official in description and thus were less informative. This was more so after 1581. But this lacuna was filled up by the narratives of J.H. Van Linschoten,[21] and Fedrici, etc.

Cesare Fedrici, an independent merchant and writer-commentator, first came to India in 1564. He stayed in the East for about eighteen years from 1564 to 1582, the longest period of stay in the East of any Europeans till then. He travelled far and wide in India and the East Indies.[22] The accounts of all the individual traders like Linschoten, Fedrici, Fitch, etc were published during the last two decades of the sixteenth century during 1587-99.

Initially, Italians and Catalans were the forerunners of sea charts called Portolans or Portolanos, but later Portuguese and draughtsmen from other countries came in this business. These charts were made by seamen for seamen, and thus were practical and utilitarian in nature. The Portolanos were usually very accurate as these incorporated a lot of practical experience and were being used by seamen regularly in their voyages and, therefore, were constantly being improved upon by adding new information by seamen as and when they got it.[23]

Portolano or Catalan sea charts as these are also called, were mostly of Mediterranean and Black seas but some very good Catalan sea charts, covering the whole sea world, mostly of the fifteenth century AD, have also come down to us. Most of these Portolanos are in atlas form, containing separate charts of various regions and sometimes in the form of a world map as well. The earliest of such atlases is dated 1320. The more famous Catalan world map is of 1375 and was prepared for Charles V of France. The outstanding feature of this world map is that its European portion is in portolano form, the central portion, including Middle East and Asia are, perhaps, from the latest information available then and the far East is based on Ptolemy's *Geographia.* The map also contains details showing areas of travels of Marco Polo.[24] The whole of Asia is shown perhaps for the first time. The map does not incorporate northern and southern regions of the world and also the southern part of Africa.

Portolan charts had also changed very much in appearance; from mere utilitarian items, they became pieces of art for the drawing rooms of the rich. But, in form and content, there was no change. No longer were the interiors of the land and other blank spaces in the sea seen on such charts. These began to be filled in with decorative, pictorial and heraldic motifs, full of fantasy and legend.

In the sixteenth century the Portuguese were considered the greatest makers of Portolano charts. This was, perhaps, because of the fact that they were in the forefront of finding a sea route to India, and were eager to grasp the opportunity to monopolise spice trade. With the successful voyage of Vasco da Gama, in the fifteenth century, the Portuguese authorities, who were financing these voyages, wanted to keep the details a state secret. To this effect King Manuel issued a decree on 13 November 1504 prohibiting the graphic representation of the routes followed by their ships under pain of death. So the routiers, charts and maps giving the details of such journeys became state secrets. To keep thier information always up-to-date, a standard map of the world, according to the latest information, was kept in the *Armazens da Guine e India.* This was referred to as the official *padrao.* All the new discoveries, as soon as these were available, were recorded on the *padrao.* Whenever an expedition was launched by Portuguese, the Captains of ships were provided with meticulously drawn copies from the *padrao* and were used to be redeposited on return.[25] From the available information on the suppression, there is evidence, that despite strict censorship, in the first half of the sixteenth century, the monopoly on news was very effective but this rigid system of control broke down by the middle of the century. Copies of the Catalan charts and general maps along with other information began to be smuggled out in many ways, mentioned earlier, and were being used by navigators and cartographers at large.

We have seen the importance and impact of Ptolemy's *Geographia* till the end of the first millennium of the Christian era. Greek and Latin manuscript copies of the *Geographia* began to circulate in western Europe in the fifteenth century. There were many knotty problems posed by the original *Geographia* of Ptolemy, as also the later editions of the text as well as the maps. The maps were no doubt a later addition, probably of thirteenth century, incorporating data from Arab and Byzantine sources, for example, the geography of individual countries as shown on the world map, and the landlocked Indian Ocean, etc. On deeper study it would be clear that such information was not commensurate with the original text.

During the next few centuries (eleventh to fourteenth) not much cartographic activity is noticed.

A manuscript copy of Ptolemy's *Geographia* was found in the collection of famous Byzantine collector, a monk, Maximos Planudes (1260-1310) which is now in the Vatican Library, but no maps are incorporated with this copy. This manuscript copy must have been studied by Agathodaemon, another cartographer of that time, as the remarks made by him regarding the construction of his world map were appended in the margins of the copy of the *Geographia* belonging to the monk. This map was in a single sheet and also 26 other regional maps were compiled by him, as per the remarks. Other manuscript copies have also been located. One was with Nikephoras Gregoras (1295-1359), also a Byzantine scholar. In this copy also, Gregoras had commented in the margin and himself prepared a world map in four sheets as well. The monk Maximos Planudes (1260-1310) had stated clearly that he prepared his maps on the basis of information contained in Ptolemy's *Geographia* which was in eight books or volumes. This confirms the assumption that *Geographia* was put together in the early period of the fourteenth century.[26]

It was in the fifteenth century that scholars began a critical study of the theories of map-making propounded by Ptolemy. His *Geographia,* along with the maps attached therewith, also came under greater scrutiny.

Barely, two decades after the publication of the Latin version (1406) of Ptolemy's *Geographia,* contemporary geographers and cartographers began to feel that it was incomplete or inadequate. In 1427, in the manuscript copy of *Geographia* which was prepared by him Guillanme Fillastre incorporated in his world map, Claudius Clavus's map of the countries of the North.

There is another copy of *Geographia* prepared in 1446 by a German cosmographer, Donnus Nicolaus Germanus. This is preserved in Este Library at Modena, an island near Venice. He redrew all the regional maps of *Geographia* on a new trapazoid projection. His maps gave a completely new look—mountains, rivers, lakes and even boundaries were shown differently; figures for latitude and longitude were placed not on them but between them; towns whose astronomical coordinates were known and those with their coordinates not known, had different signs to denote them, etc. This atlas, incidentally, was the basis of the first printed edition of the atlas, printed at Bologna in 1477,[27] after the art of printing was introduced in Europe in 1450.

Soon after, Nicolaus completed his first manuscript copy. In this manuscript edition, the world map was redrawn by him on a new projection, and to it, he added, as had been done earlier by Claudius, maps of Scandinavia and other countries of the North. He also incorporated three new regional maps to the atlas.

Just two years after this Nicolaus brought out his third edition in 1468. This was necessary to correct the location of some islands in the north region like Greenland and Iceland on the basis of the new information available. He further added two more regional maps. This atlas was reprinted in 1482 and 1486 at Ulm, and these are known as Ulm editions.

Further changes in the original atlas were made by Pietro del Massaio, in his four manuscripts copies made in 1469, 1472 and two others that are not dated. One of these was signed and the other was anonymous, but they are generally ascribed to him. In these copies, a great number of mountains, rivers that were not included by Ptolemy, have been added and named on the world map. Again, places astronomically fixed and others not so determined, had been shown differently, as also the boundaries. The striking feature of these copies was that the place names, etc., were written in both Greek and Latin.[28]

Unique translations of *Geographia*, in two manuscript copies, made in *c.* 1480, in verse, in Italian, have also come down to us.

In yet another manuscript edition of *Geographia*, prepared by another German cartographer, Henricus Martellus Germanus, at the end of the fifteenth century, 12 new maps were added to the original 27 maps.[29]

In further editions in 1507, reprinted in 1508, new additions were made by Marcus Beneventanus; a new world map was made by Bernardus Sylvanus in 1511 for his copy of the atlas. To the Strassburg edition of 1513, and the Basle edition of 1540 which had 48 maps, the additions were made by the famous cartographers M. Waldseemuller and Sebastian Munster respectively. Then came the editions by J. Gastaldi in 1548, by G. Ruscelli in 1561, and by G.A. Magini in 1596. The unique feature of these copies is that *Geographia* is divided in two parts, separating the old maps from the new.[30]

After the Munster's copy of *Geographia* came in circulation in Europe, the interest of western Europe in cartography, in general, began to diminish. Many of the new editions that came thereafter did not contain any new maps.

The only notable volume that came out, interestingly, was a fine edition of *Geographia* published by Gerard Mercator with the original maps of Ptolemy re-engraved, in 1578.[31]

D.F. Lach was of the opinion that L. Renou's critical study of Book IV, Chapters I to IV, of Ptolemy's *Geographia* published in 1925 was the best on the subject. No other satisfactory critical edition of the whole of *Geographia* has come out so far.[32]

Generally speaking, the maps of the fourteenth and fifteenth centuries reflect the cartographic thought of the past including the Ptolemaic ideas and the image painted by Marco Polo, Nicolo de Conti and other successive travellers of Asia and the Far East. The errors till then visible in the maps and other cartographic studies could be erased only by the voyages and discoveries of the sixteenth century which finally removed the oceanic barriers between Europe and Asia. Whatever fanciful ideas of the past inherited from Ptolemy and other Greek, Roman and Islamic cartographers of the west and the Middle East, these could be finally dropped in the face of rigid, empirical data available only during the eighteenth century.

We have seen that during the fifteenth century Ptolemy's *Geographia* created almost a stir in western Europe and began to be widely copied. Most of the individual maps prepared during this period were either based on Ptolemy or had information from Catalan charts and maps or had bits of both. *Geographia*, its text written in the second century, and the maps, which were later additions to the texts, probably in the thirteenth century, posed many problems to serious scholars. The individual countries, the southern-land-bridge and the landlocked Indian Ocean, shown on the earlier maps, were being hotly discussed in the light of later discoveries and accounts of travellers that came to be known in Europe in the late fifteenth and the sixteenth centuries.

A world map combining the qualities of a Catalan map and a Ptolemaic one, was prepared by Pirrus de Noha. This map and the manuscript which it accompanied, both were undated. However, the year 1414 could be ascribed to the map, as both the map and the accompanying manuscript were bound together with another manuscript which was dated 1414. The manuscript was on cosmography and was written by Pomponius Mela. The European portion of the map, more or less, shows the influence of Catalan sea charts and Asia and the Far East seem to be based on Ptolemaic ideas and maps. It shows many rivers and mountains but very few place names in the Asian region. Indian ocean is shown as landlocked and the southern land mass is contiguous to Africa.

Many other Catalan charts and maps after the famous

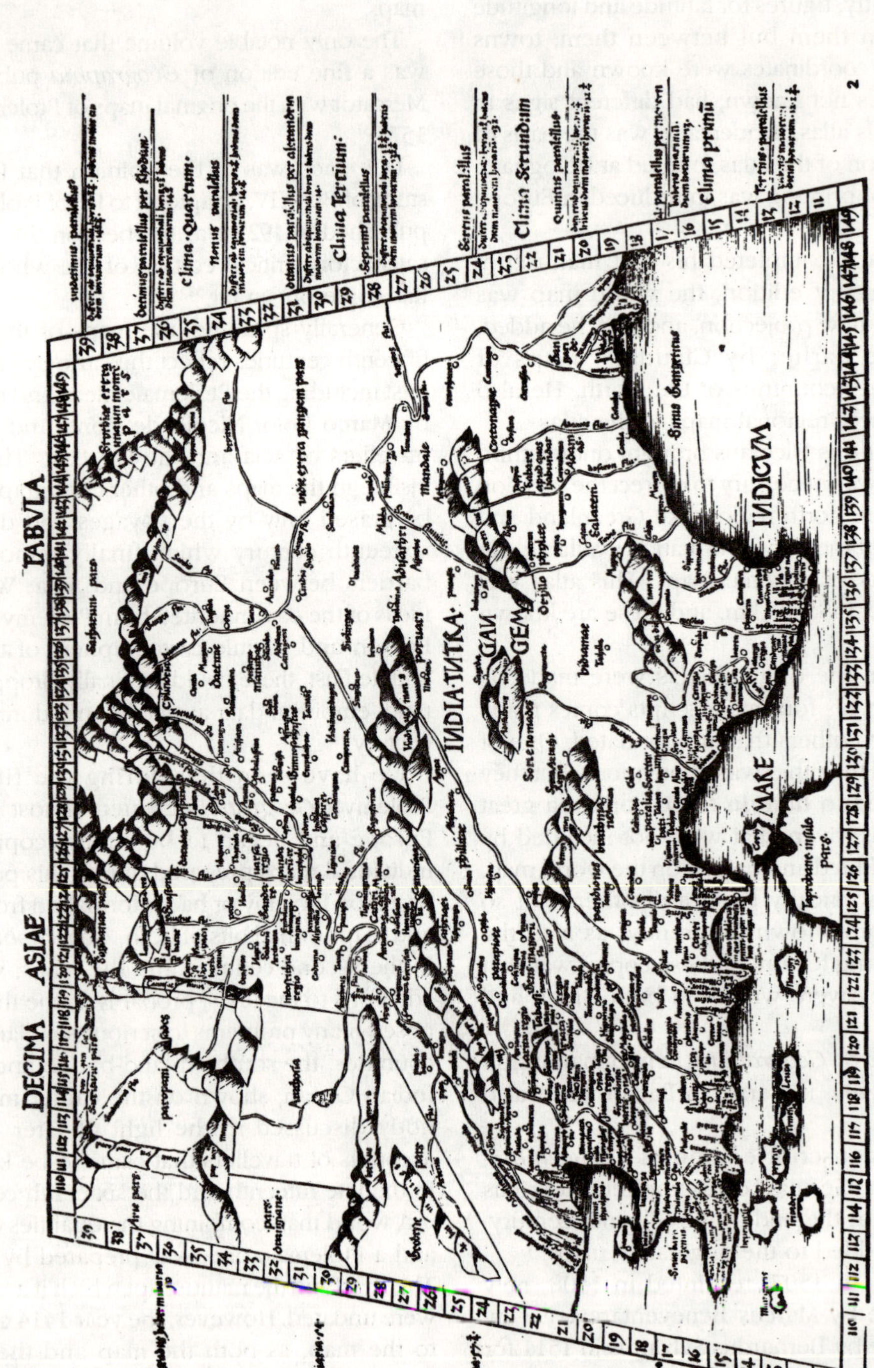

Map 8: *Decima Asiae Tabula* by A.D. Waldseemuller (AD 1513). (Courtesy: Susan Gole, *A Series of Early Printed Maps of India in Facsimile*, New Delhi 1984)

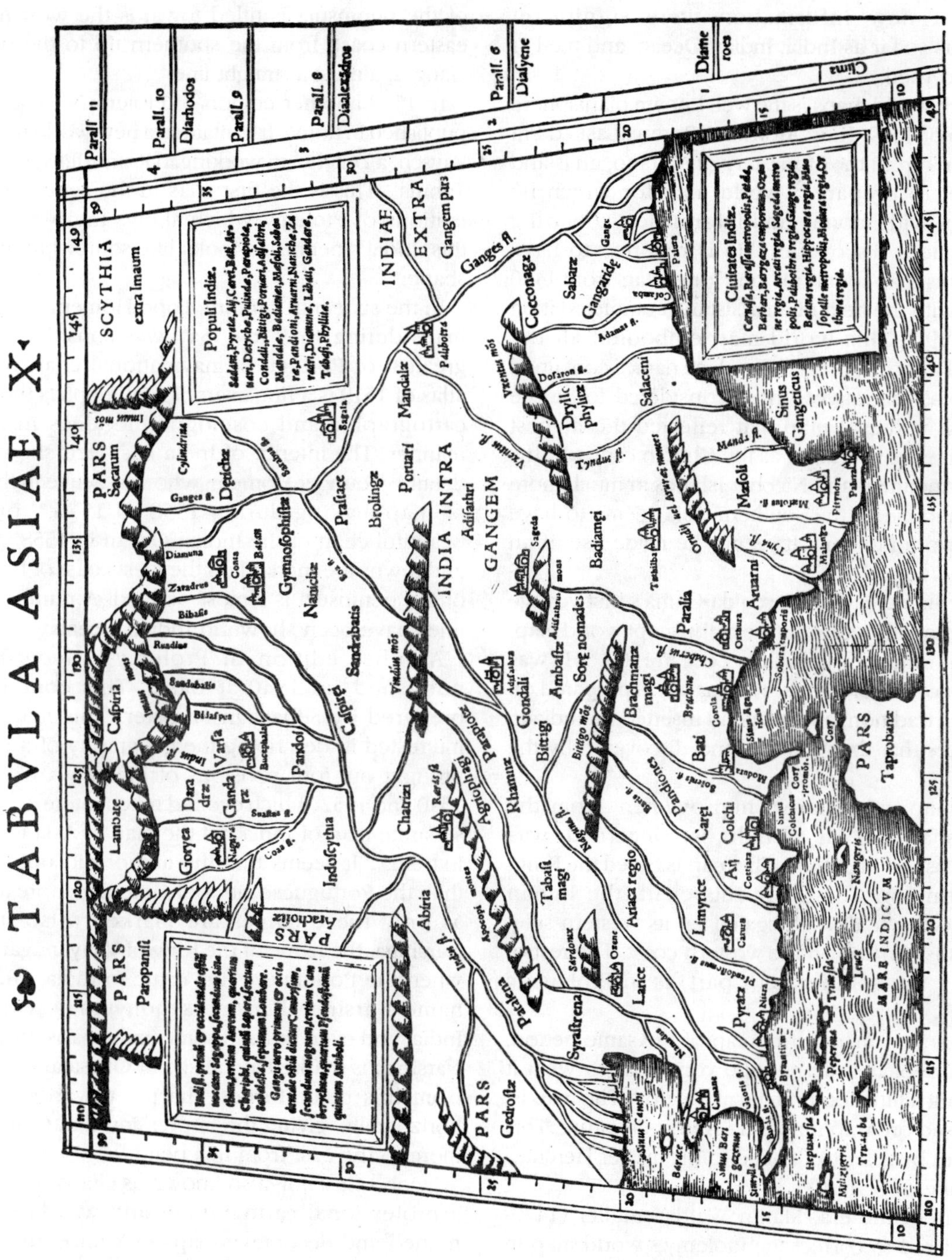

Map 9: *Tabula Asiae X* by Sebastian Munster (AD 1540). (Courtesy: Susan Gole, *A Series of Early Printed Maps of India in Facsimile*, New Delhi, 1984).

Catalan atlas of 1375, which depicts India as a peninsula, were published. These world maps were somewhat better than the circular world Catalan map of 1450, discussed earlier, now lodged with the Biblioteca Estence in Modena, near Venice.

Most of the world maps prepared during this period on the basis of new information were significant improvements so far as India, Indian Ocean, and the Far East are concerned.

One of such world maps is the well known planisphere of Fra Mauro of 1459. King Afonso of Portugal asked Fra Mauro, a monk from the Monastery of Murano, an island near Venice, to prepare a world map for which he supplied him with money and information about the latest Portuguese discoveries. He is reported to have worked on this map for about nine long years from 1450 to 1459 and with the help of an assistant, completed it on 24 April 1459.[33] This world map embodied all the geographical information and the place names contained in Marco Polo's narratives.[34] It is considered to be the latest typical medieval map that reflected the highest point in what is now known as the Church cartography. He showed Andaman and Nicobar Islands in much more realistic positions but the southern Asia seems to have been taken from Ptolemy. It seems, he made use of an Arabic *periplus* also.[35]

As already stated earlier, Vasco da Gama's first voyage was recorded on his return on the manuscript world map. This map was drawn by Juan de la Cosa in 1500.[36] He was the former pilot of Columbus. This map was prepared on the Ptolemaic tradition, and a legend inserted on India in the map states that this was the land discovered by the Portuguese.

Another map, with partial improvement from the Ptolemaic shape so far as India is concerned, was prepared in Lisbon in *c.* 1502. The map is called the King-Hamy-Huntington Chart.[37] In this chart, the Indian peninsula is shown a bit elongated on its western side. Only Calicut is marked on the western coast. The rest of Asia, mostly the southeastern part, is in Ptolemaic tradition.

There is another manuscript map of the same period, 1502, which depicts India's shape very close to what it should be, but a little too pointed at the tip.[38] The map is, perhaps, based on Arab maps and Arabic literature. This map was sent by Alberto Cantino to his master Hercules d'Estate, Duke of Ferrara.

Like others of the era, Martin Waldseemuller (1470-1518) also dared to correct the Ptolemy's world map in the early years of the sixteenth century. It formed part of 1513 edition of Ptolemy and continued to be considered the standard atlas for more than half a century until 1570 when *Theatrum* by Ortelius came out.[39] M. Waldseemuller became interested in new discoveries early in his cartographic career. Just about nine years later in 1516 a new world map was made by him. How the new informations brought in by discoveries and voyages brought a change in him, is surprising. Ptolemy was almost entirely discarded, and his place was taken by Portuguese and Spanish discoveries. Indian peninsula is shown here in its triangular form, although the lower half of the peninsula is titled towards the west making the eastern coast, from the southern tip to the delta of the Ganges, almost a straight line.

In 1508 another edition of Ptolemy's *Geographia* was published in Rome. It contained a new world map by Johan Raysch, a Dutchman working in Rome, Italy, entitled 'Nova Tabula'. This is the first atlas in the series of published editions of Ptolemy which shows the New World.[40] In this, India's peninsula looks like somewhat in its proper shape.

In the sixteenth century, Lopo Homem who produced maps during 1519 to 1554[41] was considered to be the greatest of Portuguese navigational chart makers. His atlas of 1519 is a fine example and depicts the prevailing cartographic and cosmographic ideas in a masterly manner. The interior of India has been shown in much detail.[42] Younger Homem, who was active in the business of map-making during 1558 to 1576,[43] produced a beautiful chart of the Indian Ocean in 1558. Many ports and towns are marked on the west coast of India. The tip of the peninsula is almost pointed. A number of rhumb lines have been shown in Indian Ocean.

Another edition of Ptolemy's *Geographia* was published in 1540 at Basle. This contained maps prepared by Sebastian Munster. This man was more interested in describing the geography of a country. He brought out four revisions of Ptolemy's *Geographia* in 1540 and 1552, which evoked much interest.[44] There is no separate map of India but the map of Asia is titled 'India Extrema'. It seems that the author did not know more than the Portuguese cities of Goa, Cananore and Calicut, as only these names are marked. Also included is Narsinga, the name of the King of Vijayanagar at the time when the Portuguese first came to India. Gradually the name Narsinga became synonymous with southern India, and even India in some early maps. In this edition, Narsinga is marked at the site of Orissa and the coastline from this place turned sharply towards east, almost horizontally, giving the impression that Ganges' delta is more to the east from this bend.[45]

Jacobo Gastaldi, also known as Giacomo Gastaldi was a professional cartographer and was famous for his refined and decorative maps in Venice, in Italy, during the mid-sixteenth century,[46] especially between 1548 and 1574. At this time, Rome and Venice were prominent map-making and map-publishing centres. Many individual and regional maps came out in this period.[47] In 1548, in yet another edition of Ptolemy, Gastaldi, along with other maps, also contributed a separate new map of

Map 10: *India Extrema XXIII Nova Tabula* by Sebastian Munster (AD 1540). (Courtesy: Susan Gole, *A Series of Early Printed Maps of India Facsimile,* New Delhi, 1984).

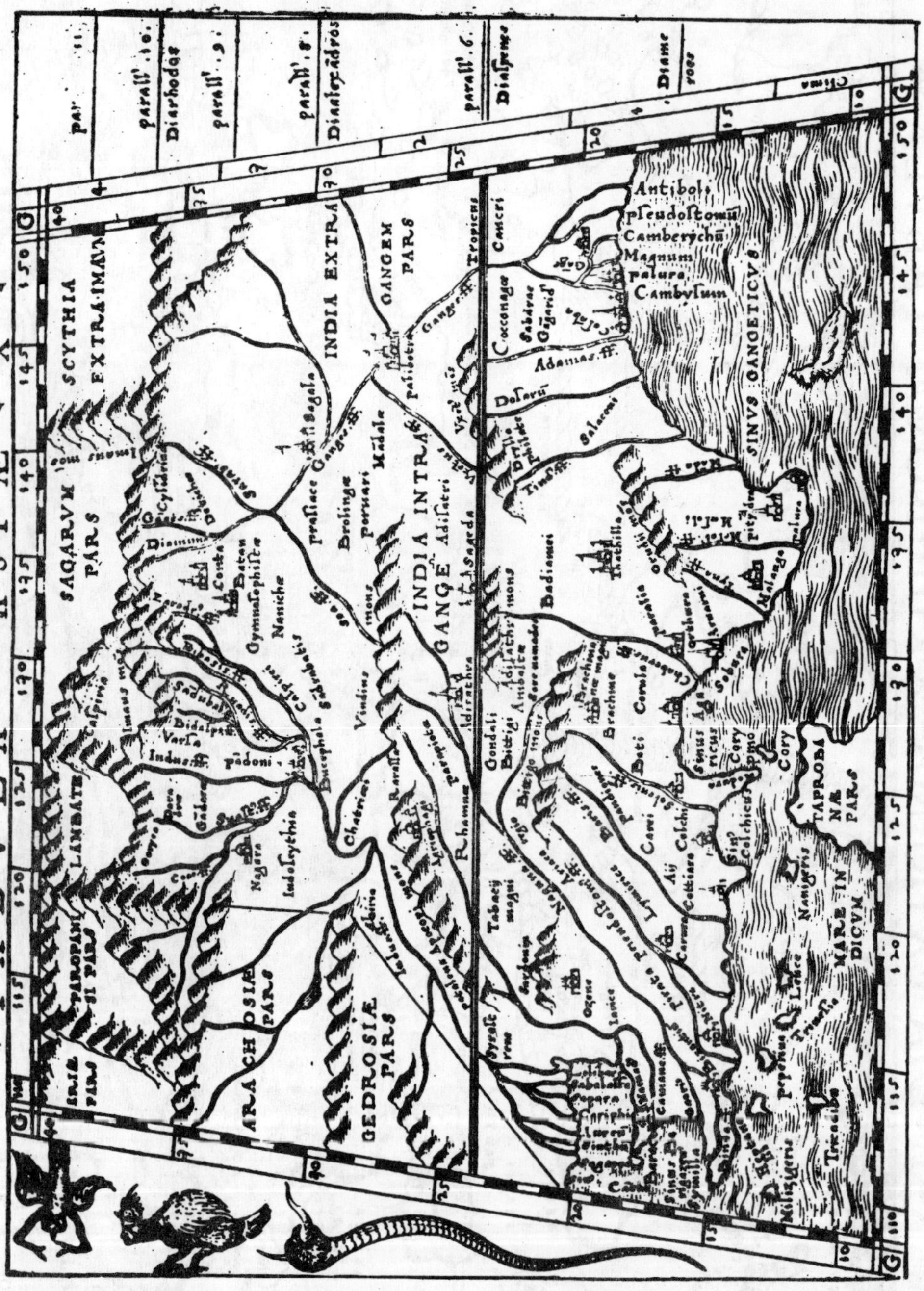

Map 11: *Tabula Asiae X* by Jacobo Gastaldi (AD 1548). (Courtesy: Susan Gole, *A Series of Early Printed Maps of India in Facsimile*, New Delhi, 1984).

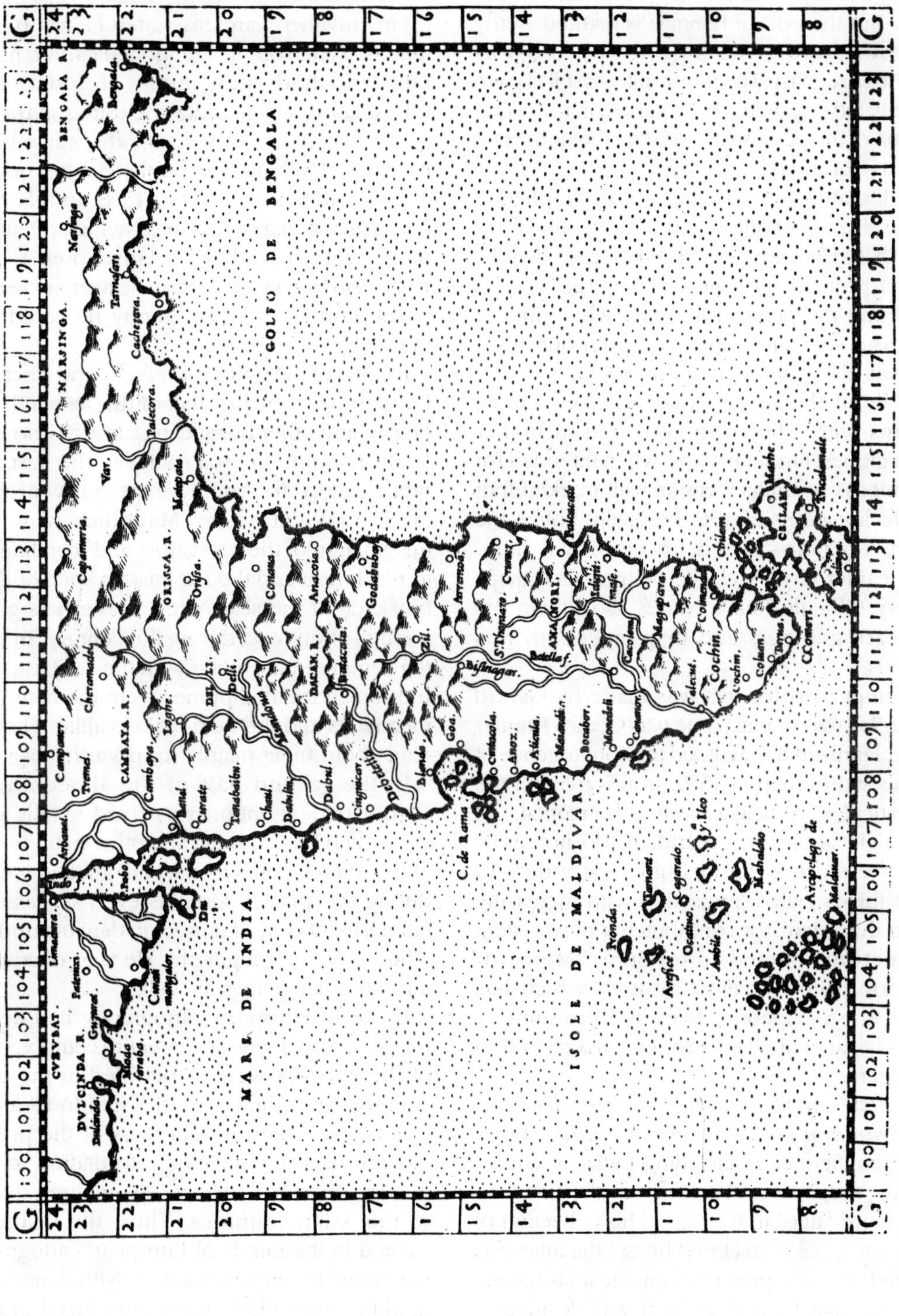

Map 12: *Calecut Nuova Tavola* by Jacobo Gastaldi (AD 1564). (Courtesy: Susan Gole, *A Series of Early Printed Maps of India in Facsimile*, New Delhi, 1984).

India on a full page.[48] This was, perhaps, for the first time that such a prominence was given to India individually, but the map was entitled simply as *Calecut.* The map shows the peninsula of India only from a little west of the Indus delta to the mouth of the Ganges, but no delta, as such, is shown, nor the river Ganges which falls in the Bay of Bengal. But the port of Bangala is marked. Many more place names are given on the west coast as compared to the east coast. The peninsula is itself shown as quite an elongated one, with Delhi marked in the centre of the top margin of the peninsula almost on the horizontal line with Bombay of today.

Much travel literature was coming out in the market during his time. He also contributed a reduction of his world map in 1548. His map of 1561 Asia is considered a fine example of professional map-making. This map is south-oriented and the peninsula is somewhat blunt at the tip. India, as a separate map sheet taken out of his map of Asia, was also published about this time.

Till the middle of the sixteenth century, on general maps meridians were shown to converge in higher latitudes, as on globes. On such maps, when intended for navigators, compass bearings were as well marked, but no magnetic declineations used to be given, unlike the modern surveyed maps based on empirical data. Navigation, in earlier days, with the help of such maps used to be very difficult. Pedro Nunes had drawn the attention of cartographers and map-makers to this problem in 1530s.[49]

The first attempt to solve this was made by Gerard Mercator (1512-94) whose real name was Gerard Kremer which was later latinized. He avoided the convergence of meridians in his world map in 1569. In this map all the lines of the graticules, vertical as well as horizontal, cut each other at right angles, i.e. meridians do not show curvature of the earth even in higher latitudes. In technical term this is called 'projection with increasing latitudes'. With this, Mercator contributed a new system of map projection, named, after him, 'Mercator Projection' during the sixteenth century, which has revolutionised the navigation. Even the present-day hydrographic navigational charts use this projection to a considerable extent.

The map entitled *Nova et accurate orbis terrae descriptio ad usum navigantium emendate accommodata* shows southern Asian configuration, including India, in almost the present-day perspective. Mercator was the first man who used the word 'atlas' in the title of his collection of maps. Unfortunately, he died (1595) before the atlas was finally published. It was a grand design, the atlas having 100 maps. It was visualised in 1578. It was decided to bring them out in three parts. The first part came out in 1585, the second in 1589 and by the time the third was taken up, he died. It was brought out by his son, Rumold in 1595. In the same year, all the three parts were also brought together and thus the final atlas was issued in 1595.[50] Mercator's maps that were part of the atlas, were original, drawn and partly engraved by himself.[51]

Jan Huyghen van Linschoten (1563-1611), was a very ambitious person. As mentioned earlier, he had stayed at Goa and had seen much of India. He collected information about India including maps, charts, pilot books, etc. from every source that came his way. On his return to Holland, he published his book *Itinerario, Voyage often Schipvaert* in 1598. In this book, he even made use of maps prepared by Pieter Plancius. Three of these were included by Linschoten to illustrate his *itinerario*.[52] In his *itinerario* he also made use of the best of the Portuguese cartography he could lay his hands upon.

In spite of the policy of secrecy and strict control by the Portuguese in the first half of the sixteenth century, Italy, France and other Europeans came to know of the details about Asia and the sea routes to the East. Many nations were eager to know, especially after the grand circumnavigation by Magellan which stirred the imagination of the navigators and merchants alike—how to reach Asia and the Far East, in spite of the Portuguese blockade of the sea routes. Westerners gradually learnt enough about Asia during the century. In the first half of the century India, along with the Far East, especially East Indies and the Philippines were the main interest of the merchants and the map-makers alike. They were able to read about India mainly from Varthema's account.

In the second half of the sixteenth century, the Europeans's attention was fixed mainly on China and Japan, and less on India and other nations of the Far East. Many travel collections and atlases were published during this period. Ortelius even surpassed Ptolemy in fame. His atlas was far more in demand and as such many editions were brought out. He was, of course, translated into other languages as well.

By 1600, European cartography had developed a set image of Asia. In the seventeenth century, only the details of the interiors were to be filled in the maps of Asian countries including countries like India and China. This was because the cartographers of the previous century did not have the specific information to complete the picture. They had mostly reconnoitred the coastal areas of the Asian countries. Thus, the image of Asia that existed in the minds of European cartographers in 1600 continued to remain as it was, with minor alterations and modifications which were introduced gradually until it was revolutionised in the eighteenth century.

Map 13: Detail of *India Orientalis* by Jan Huygen van Linschoten (AD 1596). (Courtesy: Susan Gole, *A Series of Early Printed Maps of India in Facsimile,* New Delhi, 1984).

NOTES

1. D.F. Lach, op. cit., Vol. I, p. 50.
2. D.C. Sircar, op. cit., p. 250.
3. Leo Bagrow, op. cit., p. 58.
4. D.F. Lach, op. cit., Vol. I, p. 50.
5. Ibid., p. 51
6. Ibid., pp. 34-5.
7. Ibid., p. 36.
8. J.H. Parry, op. cit., p. 7.
9. Ibid., p. 9.
10. D.F. Lach, op. cit., Vol. I, p. 59.
11. Ibid., p. 61.
12. Ibid., p. 63.
13. J.H. Parry, op. cit., pp. 138-9.
14. D.F. Lach, op. cit., Vol. I, pp. 164-5.
15. J.H. Parry, op. cit., p. 193.
16. D.F. Lach, op. cit., Vol. I, p. 152.
17. Ibid., p. 175.
18. Ibid., p. 210.
19. Ibid., p. 490.
20. J.H. Parry, op. cit., p. 193.
21. Ibid., p. 194.
22. D.F. Lach, op. cit., Vol. I, p. 490.
23. R.V. Tooley, op. cit., p. 15.
24. Ibid., p. 15.
25. D.F. Lach, op. cit., Vol. I, pp. 218-19.
26. Leo Bagrow, op. cit., p. 36.
27. Ibid., p. 79.
28. Ibid., p. 80.
29. Ibid., p. 81.
30. Ibid., p. 86.
31. Ibid., pp. 77-86.
32. D.F. Lach, op. cit., Vol. I, p. 68.
33. Leo Bagrow, op. cit., p. 72.
34. R.H. Phillimore, *Review of Imago Mundi,* Vol.XII, 1956, ed. Leo Bagrow. The author possesses the signed Ms.
35. Leo Bagrow, op. cit., p. 69
36. Leo Bagrow, op. cit., p. 107.
37. D.F. Lach, op. cit., Vol. I, p. 219.
38. Leo Bagrow, op. cit., p. 108.
39. D.F. Lach, op. cit., Vol. I, p. 220.
40. Susan Gole, *India within the Ganges*, New Delhi, 1983, p. 42.
41. Leo Bagrow, op. cit., p. 113.
42. D.F. Lach, op. cit., Vol. I, p. 220.
43. Leo Bagrow, op. cit., p. 113.
44. D.F. Lach, op. cit., Vol. I, p. 222.
45. Susan Gole, *India within*, etc., op. cit., p. 46; Map - 1540, *Munster's edition of Ptolemy.*
46. Leo Bagrow, op. cit., p. 135.
47. D.F. Lach, op. cit., Vol. I, p. 222.
48. Susan Gole, *India within*, etc., op. cit., pp. 46-7.
49. Leo Bagrow, op. cit., p. 118.
50. Ibid., p. 180.
51. Public Archives of Canada, National Map Collection, 'Introduction', *Atlases Published in the Netherlands in the Rare Atlas Collection.* Provisional series, No. I, compiled and edited by Lou Sebock, Ottawa, 1973, p. XI.
52. D.F. Lach, op. cit., Vol. I, p. 225.

5. Age of Romance
1600 TO 1800

During the seventeenth century the Indian Ocean witnessed very hectic maritime activity. There were many individual European travellers as well as officially sponsored and encouraged companies of traders whose vessels moved to and from India. It was during this time that individual travellers dared to venture into the interior of India. Their accounts, when published, generated great interest in the general public in Europe. As a result there was a great demand for books containing details of travel accounts and the latest maps of the regions showing the discoveries of strange lands and their peoples. Some publishers were excellent map-makers as well, and were interested in map-making and engraving.

On the formation of the East India Company in 1600, the English began participating in maritime trade activity in the eastern waters. They sailed on their first official voyage in 1608. William Hawkins arrived at Surat in that year. He met Jahangir, the Mughal Emperor, at Agra. The influence of the Portuguese and the Dutch at the Court immediately started working against the interest of the English. Jahangir did not immediately grant trading facilities to the English. Hawkins waited for a long time and finally returned to England in 1612 after establishing a small establishment at Surat. By the end of the year, circumstances moved in favour of the English. An English naval force under Captain Thomas Best defeated the Portuguese navy, in a fierce naval battle, near the entrance of the Tapti river which was blocked by the Portuguese, stopping the English from entering it. Soon after, in another naval engagement in December 1614, the English finally shattered the influence of the Portuguese in India. After getting a foothold on the mainland in 1612, the English gradually established commercial ties with the subordinate local agencies, which offered them opportunities to move inland. They, were thus, able to establish settlements at Ahmedabad, Burhanpur and even in the heart of Mughal dominions at Ajmer and Agra. The English factory at Surat, henceforth, became the chief English settlement in the East. For about half a century, the English and the Portuguese fought each other in Europe as well. At last in 1654, the Portuguese recognised the full rights of the English to the eastern trade in India and the Far East.

In the meanwhile, in 1615 the Company requested the King James I to send an ambassador to Emperor Jahangir with the intention of obtaining official permission to establish factories in India, as the trading stations were then called in the Company's trading parlance. Sir Thomas Roe was consequently sent to India. He met Jahangir at Ajmer, in Rajasthan, where he was camping at the time. Roe was able to obtain permission for the establishment of factories at certain towns. He had also the good fortune to accompany the Emperor on one of his military expeditions to the north-western regions of India. Finally, he left for England in 1619 on board the ship *Anne*. The first mate of the ship, William Baffin, was also a cartographer in his own right. He heard from Roe, en route to England, the details of his travels in India and also obtained as much geographical information as was possible.[1] William Baffin, later in the year, published in London his map of the Mughal Empire, mostly based on the information supplied by Sir Thomas Roe.

As already mentioned, during the seventeenth century, the main interest of cartographers shifted from the coasts of India to the interiors of the country. As a result, a number of books were published during this century giving the descriptions of places that were far away from the coastal belt. These books not only contained information about the social and cultural activities of the people but also geographical details of that part of the country which the travellers visited. They also incorporated a good amount of topographical information of the surrounding regions from hearsay accounts. Thus the interior of India was gradually being thrown open to the Europeans, through literature.

Perhaps the earliest such travel book was *Voyages and Travels, a True and almost Incredible Journey*, published in 1612 by Robert Coverte, a castway on a ship, who first had been taken to Surat and who then travelled to the capital of the Mughal Empire, Agra. There he met the leader of the first official voyage of the English East India Company, William Hawkins. He travelled back to England via the port of Ormuz.[2]

In 1614, Mendes Pinto published his account of the journey to the supposed place of origin of the holy river Ganges, a lake which he called Lake Singapamor, located somewhere in the present south-eastern region of Tibet or north Burma. The 1636 map of H. Hondius also shows

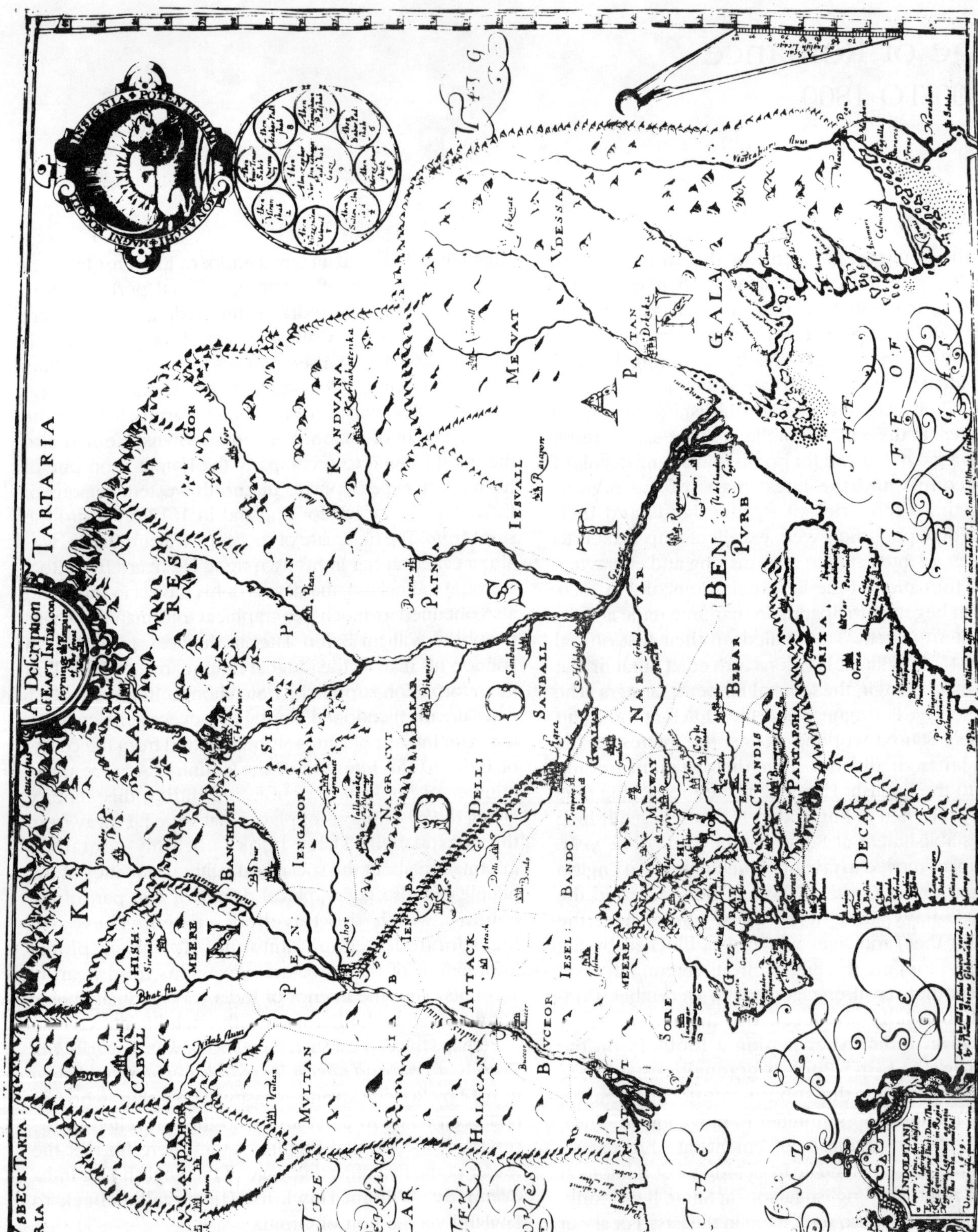

Map 14: *A Description of East India, conteyninge th' Empire of the Great Mogoll* by William Baffin (AD 1619). (Courtesy: Susan Gole, *A Series of Early Printed Maps of India in Facsimile*, New Delhi, 1984).

this lake, but like Ortelius called it Chiamay lake.[3] As per Pinto's version, four rivers emerged from this lake, one flowing east, the next running south-east, the third towards south and the fourth, which he thought to be the Ganges, flowing a little south-westward, via the region he termed as Bengala, to the Bay of Bengal. Hondius and Blaeu, both called the fourth river Caor river. This book became very popular and a number of translations were also published subsequently: the spanish translation came out in 1620, the English book was published in 1653, and in between, Samuel Purchas, the successor of Richard Hakluyt, published his abridged version of this book in 1653.[4] This description of the source of the Ganges became so popular that in almost all the maps that were published in Europe during this period—and incidentally, most of them were copies of Baffin's map—included this lake, but, interestingly, in Baffin's original map (based on the Roe's version of the geographical details), this lake is missing.[5]

In 1626 Samuel Purchas brought out his book *Purchas His Pilgrimes.*[6] This was very well received by the public. It included the story of Sir Thomas Roe, his journal and a map. It also incorporated the diary of William Methold. During his sojourn in India, Methold had heard of diamond mines of Golconda, and his account in the book by Purchas is detailed. Two similar books by travellers came out in 1644 and 1652. These were the accounts of Mandeslo and Pietro della Valle respectively. Another renowned traveller, Jean Baptiste Tavernier, lived in India for many years and made many journeys through India between 1640 and 1667. He also kept records of the distances marched by him stage by stage, along with other topographical details of rivers, mountains, etc. His account was published in 1677.[7] There were other writers who published travel books giving accounts of other travellers. These books contained maps as well. But unfortunately, the maps in these books, by then, cease to contain any new geographical discovery or information. These were mostly repetitions of early maps taken from old blocks.

In 1672, Phillipus Baldaeus, a missionary who had worked for many years in Goa published a book dealing exclusively with peninsular India for the first time in Holland.[8] The book, *Naauwkerige Beschryvinge van Malabar en Choromandel,* contained views of some of the Portuguese settlements in India along with two new maps.[9] These two maps together, cover the entire southern peninsula. It appears on examination that the map-maker had deliberately shown the vertically western half and the eastern half separately. Both the maps contained detailed information of the coastal regions and also many place names of towns and villages particularly on both the banks of the major rivers of the peninsula. But an interesting feature is that both these maps are differently oriented—map of the Malabar coast is east-oriented, i.e. east is shown on the top, and the other half, the map of the Coromandel coast, is west-oriented.

As more and more cartographers began showing details of the interior of India during this century, much confusion prevailed amongst them regarding the origin of the much revered river Ganges, and the mighty Himalayas.

As mentioned earlier, Lake Singapamor of Pinto's description of 1614 was first shown by Ortelius in his map of Asia of 1580.[10] Ortelius called it Chyamai lake. But Roe who lived in the Mughal Court for a considerable period and even accompanied Emperor Jahangir, to the Himalayan region in the north-west, had never heard of it. So Baffin could not include it in his original map. Instead, Roe's information was that the Ganges issued out of a cow's mouth near Hardwar. Somehow, the cow's mouth was changed, in some later maps to a small lake. Some of the famous cartographers of the last century, like Ortelius, Gerard Mercator, de Jode, and H. Hondius (of the early seventeenth century) in their later editions, had shown the lake of Pinto, but they did not believe that the Ganges originated from a lake, but instead were of the opinion that it issued out of a place far north in central Asia, flowed down in north-south direction with no Himalaya mountains in between the source and the sea to obstruct it or change its course.

In 1637 a Dutch cartographer made an interesting map of Asia[11] wherein two rivers both representations of the same river perhaps, from two different sources, were shown on one map, with two names given to it. This map was published by Rumold who was none other than the son of the famous cartographer, Mercator. The holy river Ganges is shown flowing almost due east and falling into the China Sea; the other river, named Guenga is depicted in the peninsular region and flows into the Bay of Bengal. Palimbotra or Patliputra is shown on the former river while Delli or Delhi is marked on the latter one.

Even in the maps of *Geographia* of Ptolemy of the second century AD, Himalaya mountains are clearly marked, but it is really surprising that Roe had not heard of them and so Baffin could not mark these mountain ranges on his map at a place where these should have been. Some writers even now maintain that as there were many Hindu holy places in the region, and as Mughal rulers were the masters, he was not told of the existence of the Himalaya by any one. This, however, does not seem to hold good when, we know, that Roe had been to the north-western Himalayan region along with the Mughal army. He must have noticed the mountain ranges and could not have failed to enquire about the same. But still, William Baffin's map did not show them at the proper place and instead, they were marked too far away to the north. Either Baffin could not place them properly or Roe, if he knew of them, could not describe the topography of the region correctly. As already pointed out earlier, the massive Himalayan ranges had not been shown by any Dutch cartographers like Ortelius, Mercator, de Jode, Hondius, etc., during the early seventeenth century.

Side by side with the new travel literature giving details of the newly discovered lands, there was a tremendous individual mapping activity seen through the seventeenth century.

The earliest important world maps of this century that aroused European interest in the new discoveries were by Father Matteo Ricci.[12]

Ricci was born in Italy and was educated in Rome. As a first Jesuit missionary, Father Ricci came to Macao in China in August 1582. He was greatly impressed with the geographical knowledge of the Chinese. Being a cartographer himself, he immediately started collecting Chinese maps and other geographical information. He lost no time in preparing his own map of the world, based on the Chinese philosophy and cosmography. This was the Chinese version of the world with China at the centre. This was perhaps also the map, that was first published in the sixteenth century in some of the editions of Abraham Ortelius' *Theatrum Orbis Terrarum*, as the world map entitled *Typus Orbis Terrarum*. Later in 1600, an enlarged and corrected edition of this map was published in Nanking. Again, a huge map, designed for a twelve-panel screen, three by two metres in size, was published in Peking on 17 August 1602. Many copies of the map were said to have been prepared. In yet another edition published in 1603, all the prefaces to the map were written by eminent Chinese scholars.

In this map the shape of the peninsula of India is shown much broader in the south and the tip is flat.[13]

Baffin's map of the Mughal Empire was printed in London in 1619 and was well received. This is the first map of the first English cartographer. The information it contained was not based solely on the details supplied by Sir Thomas Roe alone, it was supplemented from other source as well. But still, it had its own peculiarities, which have already been discussed.

In 1623, the then head of the China Mission, the Sicilian Father Nicolo Longobardi (1559-1654) and his colleague, a Portuguese, Manuel Dias (1574-1659), together constructed a large painted terrestrial globe. The peninsula of India appears to be more or less in correct shape,[14] but the tip is somewhat pointed and reaches almost the same latitude as that of the southernmost coastline of Ceylon (Sri Lanka).

In 1645, a world map published in Japan as Shoho map, unmistakably showed Ricci's influence.

In 1648, a Chinese convert, Pi Fang Chi, also a member of the society of Jesus, published a world map in Canton. The peninsular shape of India is shown as twisted—both its western coast and eastern side—are bent more to the east from the centre and above northwards. The tip of India is much elongated and bent westward, thus giving the peninsula a peculiar curved shape.

In the same year Neapolitan Father Francois Sanbiasi (1582-1649) published an oval map of the world in China. Ricci's influence again is clearly visible.

The Belgian Father Ferdinand Verbiest was another Jesuit to publish a world map in 1674. This map showed the world in two hemispheres, each hemisphere measuring 1.75 metres in diameter. The peninsula seems to be depicted, more or less, in its correct form.

A General Map of the East Indies ... by Guillaume Sanson, Geographer to the French King, drawn in 1676, was rendered into English by Richard Blome and was dedicated by him to the East India Company.[15] In this map, India's shape is very much distorted. The upper region of the peninsula of India appears to be much too tilted towards the east from its middle. In 1680, an English map-maker, Robert Morden (d. 1705)[16] prepared three miniature maps for a similar book on geography. These were: (a) Empire de Mogol, (b) India on this side Ganges and (c) The Maldives and Ceylon Islands.[17]

Frederick de Wit, one of the best known engravers of his time, came as well into publishing, and about 1688 published a map, *Magni Mogolis Imperium*. It is generally said that de Wit used to purchase old used plates of Bleau and Jansson and these used to be re-printed in his own atlases.[18]

Although map-making in Italy had declined during the seventeenth century and no new geographical information was being displayed by the geographers, some of the maps that were published during this century were more decorative in appearance than informative. Vincenzo Maria Coronelli (1650-1718), founder of the first geographical society in Italy in 1680, was a Franciscan friar and was appointed Cosmographer to the Republic of Venice. He published two maps in 1694 in the second edition of *Corso geografico universale*. The maps were: (a) Impero del Gran Mogol, and (b) Peninsola dell'Indo.[19] The second one was again included in Vol. II of *Atlante Veneto* in 1697.

In the parlance of the cartographic world, the seventeenth century is also known as 'the Century of Atlases'. Barring Ptolemy's *Geographia*, the total number of atlases that came out in the seventeenth century was much more than that in the sixteenth and eighteenth centuries.[20] This proves the general notion that during the seventeenth century the educated class in Europe took more interest in the geographical discoveries not only in the east but throughout the world, than they did in the previous and the succeeding centuries, during which time the interest was, perhaps, more regionalised.

Generally speaking an 'atlas' is a collection of maps and charts bound together. In this sense, the first printed atlas was Ptolemy's *Cosmographia*, or what is generally called *Geographia* of the second century AD, printed in 1477 at Bologna, Italy. This was continued in the sixteenth century by Lafreri (1512-77) in Rome, Duchetti (?-1585), Forlani (flourished in 1562-76), and the Bertelli (sixteenth century), all in Venice.[21] In the sixteenth century the meaning of the word 'atlas' underwent a change.

'Here in the Netherlands began to appear what we may call the first "regularly produced" atlases: collections of maps and charts of *uniform size, collected and arranged systematically* with or without accompanying text (emphasis added).' Thus Ortelius' (1527-98) 'Theatrum Orbis Terrarum' can be regarded as the world's first regularly produced atlas.[22]

This great work came out in 1570 and soon many editions were published in different languages. It contained many maps prepared by different cartographers in various sizes and scales. These were

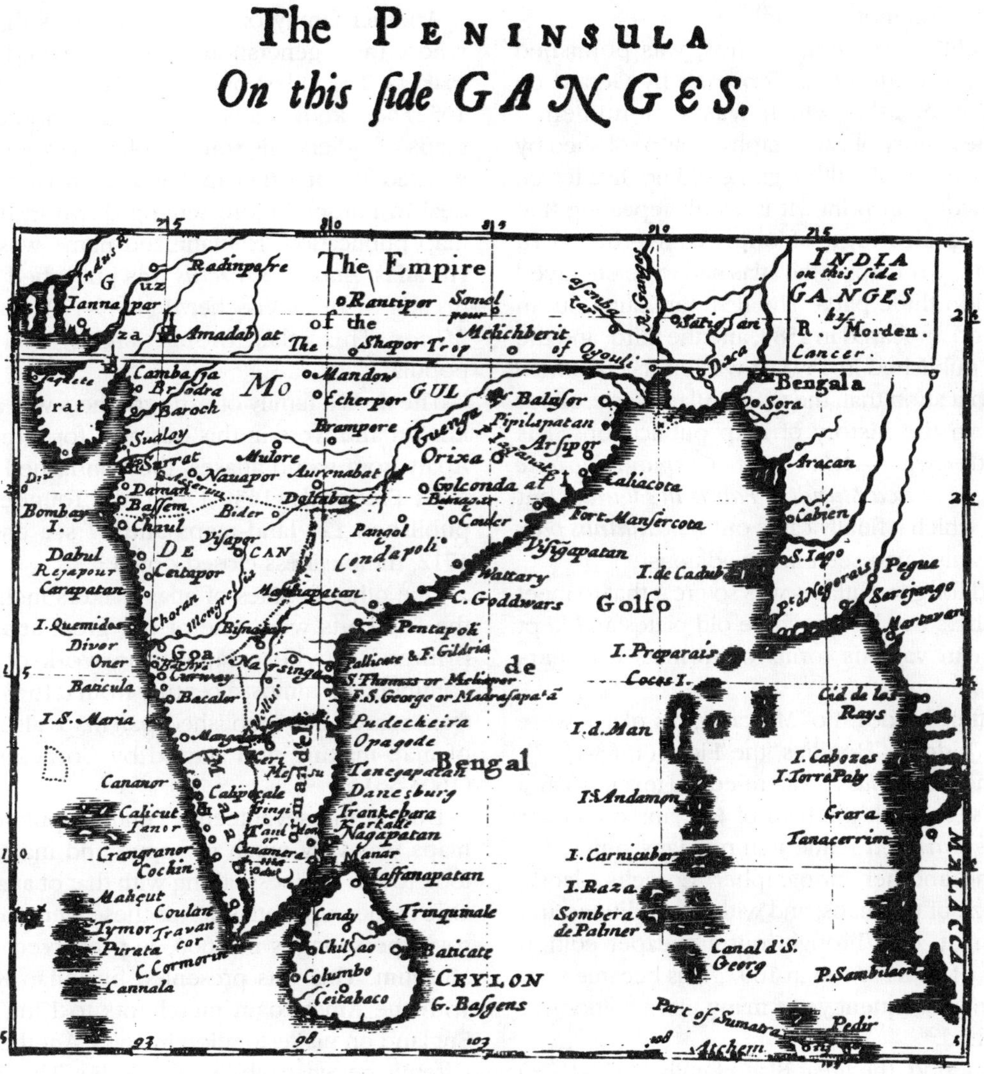

b

Map 15: *India on this side of the Ganges* by Robert Morden (AD 1680). (Courtesy: Susan Gole, *A Series of Early Printed Maps of India in Facsimile*, New Delhi, 1984).

reduced to uniform size and format. 'Another unique feature of the atlas is the "Catalogus Auctorum", a list of geographers and cartographers whose maps Ortelius used or were known to him.'[23]

Ortelius also published supplementary atlases containing new information or information correcting the previously published maps in the atlas. These atlases were called *Additamenta*. Ortelius used to collect the best available maps and reduce them to suitable size for his atlas or *Additamenta*, and each succeeding edition contained more and more maps.[24]

In 1578, eight years after Ortelius, was published another atlas *Speculum Orbis Terrarum* by Gerard de Jode. Yet another atlas which was a monumental landmark in the history of cartography, was published by Gerard Mercator in 1595 (although he did not live to see his complete works in print). It is worth repeating that because of the shortage of engravers—the work contained, in all, 107 maps—the atlas had to be engraved and published in three parts. The first part came out in 1585; the second appeared in 1589; and the third, and the final set were published in 1595. It was on the *intended* title of this publication that, the word 'Atlas' was used for the first time in the history of map publications. His proposed title was—*Atlas sive Cosmographicae Meditationes de Fabrica Mundi et Fabricati Figura*,[25] but the title under which it finally came out was *Atlantis pars Altera*....[26] This also had seen many editions.

The demand for good atlases was so great that to meet it, atlas publishers used to purchase old plates and kept printing them in various forms or formats for years together.

A decade after the death of Mercator, his plates were purchased by Jodocus Hondius, the Elder (1563-1612), and he published the *Atlas* ... with re-edited text in 1606.[27] After about a year, with the help of Jan Jansson, better known by his *latinised* name Joannes Janssonius, his son-in-law, and another cartographer, Cornelis Claesz, reduced the size of the maps, and with the additions and change of format, etc., brought out a cheaper edition under the title of *Atlas Minor* in 1607. This became very popular and in subsequent years many more editions of it were published.[28]

During this period, the great Blaeu family (1571-1673) of cartographers, father and two sons, surpassed all other cartographers of the sixteenth and seventeenth centuries. Their fame has not diminished even to this day. Father, William Janszoon Blaeu (1571-1638), also known as Guilielmus Janssonius (sometimes he used his latinised name, Caesius), his older son Joan (1596-1673), also called Joannes Janssonius of Amsterdam, and the younger son, Cornelis (*c*.1610-45), all earned name and fame as great cartographers in their own right. Blaeu, the elder, used to publish globes and maps under the name of Guil. Jansz, and to avoid confusion with the map publisher Jan Jansson, he started using his surname, Blaeu. He and his son Joan brought out their first atlas, *Appendix Theatri A. Ortelii et. Atlantis G. Mercatoris*. In 1635, another atlas of theirs, *Theatrum Orbis Terrarum, sive Atlas Novus* was published. When their father died in 1638, both the brothers carried on the business. The famous works, *Atlas Novus* and *Grand Atlas*, in six volumes, or the *Cosmographia Blaviana*, in twelve folio volumes, are the works which have as yet not been surpassed in fame.[29]

Another family of map-makers was that of Visschers whose three generations were renowned map and atlas makers. The eldest one, Claes Janszoon Visscher (1587-1652) was known a versatile engraver and used to print maps of others; his son, Nicolaes Visscher I (1618-79), worked for some time in Blaeu's firm. He travelled a great deal in Europe before settling down in the business of map publication. His latinized name was N.J. Piscator. His atlas, *Atlas Contractus* was published in 1664, while his son Nicolaes Visscher II (1649-1702) published *Atlas Minor*. This atlas, though small in size, was very popular.[30]

The de Wit family of cartographers was also quite well known and were in this business for three generations. Apart from world atlases, they published sea atlases as well. Frederick (1630-1706), the founder of the firm, published 124 land-maps and 27 sea charts. In about 1712, the business closed down.[31]

Like other families of atlas makers mentioned above, the Dankerts were one of the great atlas makers and printers, in whose family the work of map-making continued for almost a century from *c*.1633 to 1727. Apart from atlases, they published wall maps also. The business of map-making was started by Cornelius Dankerts II (1603-56).[32]

During the seventeenth century, publication of wall maps was very much in vogue and many map-makers took to this business, along with that of atlas publication. One of the 'super atlases', (as these were called) made for royal personages in 1660, is preserved in the British Museum. This was presented by Johan Klencke along with the Amsterdam merchants to King Charles II of England on his restoration in 1660. Another one is in the Offentliche Staatsbibliothek, Berlin. The atlas measures nearly six feet in height. These contain the maps made by Amsterdam map-makers.[33] One of the maps, *India Orientalis* (seventeenth century) made by Huych (Hugo) Carel Allardt, is included in the 'super atlas', preserved in Berlin. Such atlases usually contained 35 to 40 wall maps.

The practice of bringing together maps published by various publishers into one atlas (begun by Lafreri in Rome) came to its height of popularity, with the preparation in Amsterdam in 1706-10 of an atlas consisting of more than 1,400 folio sheets of maps in 19 morocco size volumes. It contained maps, plans, views, portraits, etc., of the seventeenth and early eighteenth centuries and was called *Atlas Royal*.[34]

Apart from the 'Super Atlas', normal size atlases, and bulky atlases printed in several parts, there used to be published during these centuries 'pocket' sized miniature atlases as well, based on the works of famous cartographers like Ortelius, Mercator, etc.

By the end of the seventeenth century, many European powers had established trading centres at many places on the coasts of India. They had rarely ventured into the interior even for trading purposes.

The Portuguese, as already mentioned, came first on the Indian scene. From the landing of Vasco da Gama in 1498, near Calicut, on the western coast, they took Goa in 1510; fortifications were erected at Chaul, south of Bombay in 1521; Diu, Bassein and Bombay were acquired in 1534, and Daman was taken over in 1559. Next came the Dutch, who had their settlements at Nagapatam which lies south of Tranquebar, on the eastern coast; Sadras, Pulicat, in the south and north of Madras respectively, and Bimlipatam, north of Vishakhapatnam, all were on the east coast of the peninsula; and finally, they had their trading centre at Cochin, the only centre on the west coast.

The English entered the Indian Ocean for trading in 1608. They could establish their various trading posts at Surat, Calicut, Masulipatam, Madrasapatam, etc., and had their first fortified post called Fort St. George at Madras by the end of the fourth decade of the seventeenth century. On the marriage of the King Charles II with Catherine of Braganza, Bombay was given to the English crown in 1661 as part of the dowry. The island was later handed over to the East India Company in 1669. Fort St. David, just south of Madras was acquired by purchase in 1691. In Bengal, a licence to trade in Bengal was obtained in 1650 and in 1651 the English factory at Hughly was established. In 1696 Chuttanatti, Calcutta and Govindpur villages in Bengal were similarly purchased from Azim-u-Shan, grandson of Aurangzeb. In 1658, all the settlements in Bengal and on the Coromandel coast were made subordinate to Fort St. George.

The French had in their possession Masulipatam, Pondicherry and Chandranagore since 1669, 1674 and 1674 respectively, and by 1700 the Danes also held Tranquebar and Serampore.

This shows the comparative interest being taken by the European nations till 1700 in the protection and expansion of their trade in India.

Once settled on the main land of India, they began to look around, examine the topography of the region all around their settlements, talk to the people, gather geographical information from them, and their mental horizon, in this regard, began to widen.

With the increase in trade and consequent prosperity of the Europeans, and the gradual opening of the interiors of the Indian mainland as well, maps and charts showing the new areas of the hinterlands of the port towns, were more and more in demand in Europe. Maps and atlases, therefore, continued to show various types and categories of topographical and geographical information to the curious masses. As already mentioned, atlases giving the latest maps of the countries of the world at large were being increasingly published. Along with atlases, individual sheet-maps were also being printed in great numbers. However map-making, being a time consuming and expensive process, printing of fresh maps giving the latest information, could not keep pace with the demand. Publishers were, therefore, reprinting the old maps till the time new maps became available. Sometimes, to mislead the people, they used to print, even on the old maps, legends to the effect that the maps contained the latest information available. This resulted in the mistakes in old maps being repeated year after year by many map-makers, even after fresh information was available in the newly published travel books.

The work of some of the better known map publishers during the eighteenth century can now be described. We may start with Herman Moll (d. 1732). He was a Dutchman who settled down in London in 1680. In the initial stage of his career, he was an engraver and used to engrave maps for others. Later, he became a map-maker. In 1701, he brought out his first map of India in *A System of Geography*, showing the Mughal Empire. In 1712, he brought out another map of India, *The West part of India*, in the *Atlas Geographus*, followed by yet another map that was titled *India Proper* in 1729, which was included in *Atlas Minor*.[35]

In 1706, a French publisher, Henri Abraham Chatelain (1684-1743) brought out an interesting book, a sort of short encyclopaedia, in several volumes, incorporating a small map of part of India with the unusual title *Mogol*. In 1719, the maps—*Genealogie des Empereurs Mogols Depuis Tamberlan ou Tamerlan*, with a map of Kashmir, and *Carte Nouvelle des Terres de Cucan, de Canara, de Malabar, de Madura, et de Coromandel*[36] were published, copied from the one published by Francois Bernier in 1669 in his *Atlas Historique*.

Pieter van der Aa (1659-1733), a prolific map-maker of Leiden, in his early career, drew maps to illustrate routes of any traveller who wanted his accounts of journeys to be published. The maps drawn by him were not original, and were casually drawn with little regard for geographical facts. However, his atlases, in his later years, became popular and were reprinted many times as he began to be considered a serious map-maker. In 1707, he published two maps—*Mouths of the Ganges*, and *The Kingdom of Bengale*. In 1714, in his *Atlas nouvneau et curieux* were printed *The Empire of the Great Mogol*, and *The Kingdom of Guzaratte*. Again, in 1729, he published his great work, in 66 volumes, containing 3000 sheets of

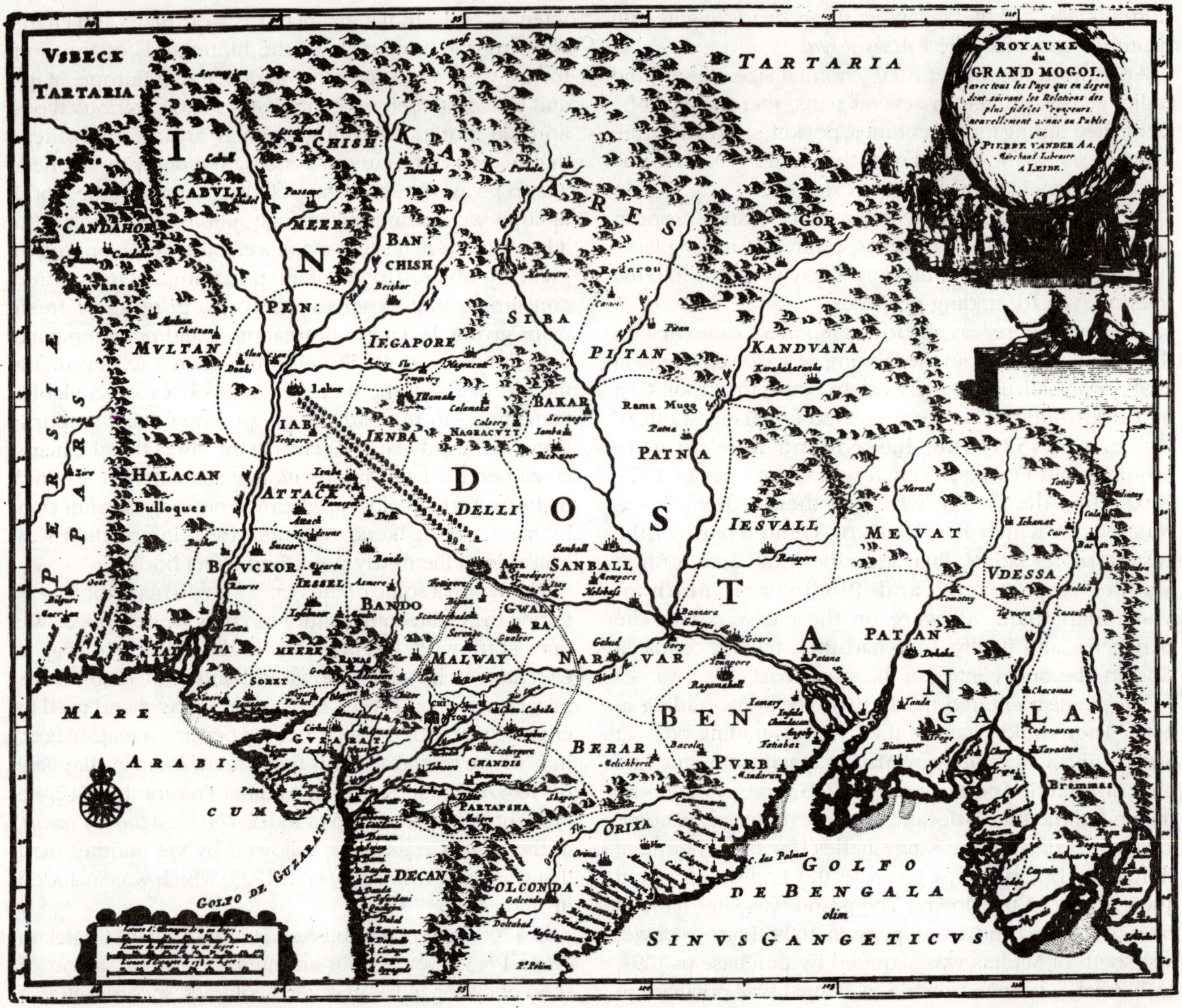

Map 16: *Royaume de Grand Mogol* by Pieter van der Aa (AD 1729). (Courtesy: Susan Gole, *A Series of Early Printed Maps of India in Facsimile*, New Delhi, 1984).

maps. Out of these, two volumes were devoted to India alone. Apart from maps, they contained city plans and views of incidents of historical importance. It appears that the map of the Mughal empire entitled—*Royaume de Grand Mogol* is a copy from that of Baffin which had been published about one hundred years earlier. But van der Aa, claimed it in the title have been 'newly given to the public'.[37]

Three interesting features, depicted in the map are worth mentioning. River Ganges is shown flowing through a 'Cow's head' and a straight road from Agra to Lahor (Lahore) in the north-west direction indicated by two rows of trees in straight lines, and the river Jemna (Jamuna) is shown issuing out from near Delli (Delhi).

In 1720, from the information collected by the famous French geographer de l'Isle, a map, entitled '*Nova Tabula Terrarum, Cucan, Canara, Malabaria, Madura & Coromandelia* was compiled and published by Gerard van Keulen (1678 1728). Gerard was the son of the famous Johanness van Keulen (1654-1715) who was well known for his marine charts and atlases. Gerard succeeded him in the business in 1704. By 1714, Gerard became hydrographer to the Dutch East India Company.[38]

The map mentioned above shows four compass roses with rhumb-lines all over the sea around the peninsular India.

In 1722, Guillaume de l'Isle (1675-1726) published a map of southern India, *Carte des Cotes de Malabar et de Coromandel*. Because of his position in the French court, he was able to procure in 1719 the latest map of southern India from the Jesuit Father I.V. Bouchet. It was later used by many others in their publications. There were other

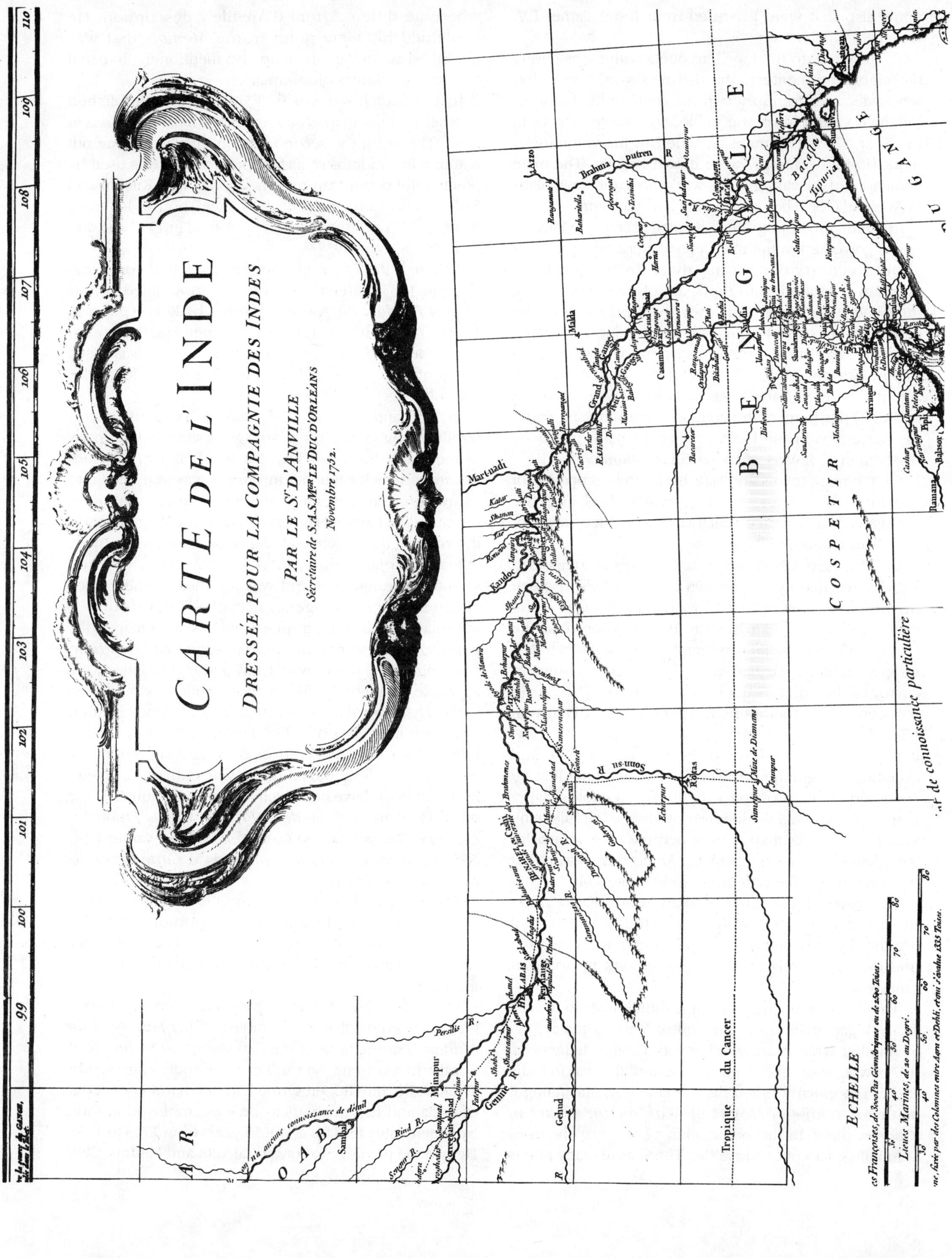

Map 17: Detail of *Carte de l'Inde* by Jean Baptiste Bourguignon d'Anville (AD 1752). (Courtesy: Susan Gole, *A Series of Early Printed Maps of India in Facsimile*, New Delhi, 1984).

maps also that were procured from Jesuit Father I.V. Bouchet.[39]

To turn now to the travel accounts with maps during the eighteenth century, the first noteworthy traveller, who was a cartographer in his own right, Francois Valentyn visited India and published his experiences in the year 1726. The maps included in his voluminous work (in five volumes) were drawn by him. The main attractions of the cartographic works were the town plans of Wingurla, St. Thomas, Hughly and Barsalore.

In Germany, during the eighteenth century, Homann family played a leading role in reviving cartography after about a century of near inactivity. In the absence of German contribution to the spate of new discoveries, they had nothing of their own to display before the nation except copies of maps being produced by other countries. The founder of the family business was Johann Baptist Homann (1663-1724). After the death of his only son in 1730, the business was taken over by H.J.M. Franz and J.B. Epersberger who continued map-making under the name of Homann Heirs. One of the maps produced by them in 1733 was *Malabar & Coromandel*.[40]

On the map, towns and forts under the occupation of the English, Danes, French, Hollanders and the Portuguese have been distinctly marked with symbols of forts with respective flags flying over them.

In 1752, there come out a map of India accompanied with an explanatory text, known as *Eclairissement* which thrilled the cartographic world of Europe. The cartographer was Jean Baptiste Bourguignon d'Anville (1697-1782), an official cartographer for the French East India Company. He had collected a huge amount of cartographic material including maps. This large collection was acquired by King Louis XVI of France in 1770. The King was most interested in Asian geography. d'Anville was considered to be a most scientific and accurate cartographer. Almost at the same time that he compiled his map of India, he prepared another large map—*Coromandel*—but it was published the following year in 1753. This map at once became very popular as it explained, to a great extent, the Anglo-French wars that were waging in the peninsular India since 1746.[41]

The map of India contains, apart from deltas of rivers in Aracan in Burma, soundings in the Bay of Bengal around the area where the river Hughly meets the sea and the adjacent islands, along the coast up to Point Palmyras.

The Coromandel map contains rhumb-lines in the sea from Negapatnam up to the port of Masulipatnam.

Another writer, cartographer was Thomas Jefferys (d. 1771) who wrote the English version of the war in India in 1754, popularly called the Carnatic war, illustrating it with his map—*The Seat of War on the Coast of Choromandel*. In the preparation of this map he used d'Anville's map of Coromandel. There were many points where he differed from d'Anville's description. He explained his view point in the *Memoir* that was published along with the map also mentioning, in detail therein, his sources of information.

Just before his death in 1771, Jefferys published another popular map, *The East Indies with the Roads* in 1768. This map, the second edition of which came out within a few months of its first publication, was used by other well-known map-makers like Thomas Kitchin and Faden (who came into business after Thomas Jefferys's death in 1771) for their atlases published in 1777 and and in 1778 respectively.

Although the Copyright Act for maps had come into force in 1735, Jefferys' map of 1768, now titled *A New Map of Hindoostan*, was re-engraved by Robert Sayer in 1789. He deleted Jefferys' name and, instead, acknowledged Rennell's surveys, as if the map was based on Rennell's maps of India of 1782 and 1788, with impunity.[42]

In 1755 and again in 1756, Antoine Francois Prevost published his collection of voyages incorporating many of the maps drawn for Jean Nicolas Bellin's *Petit Atlas Maritime*, which came out in 1763. It was translated into many European languages and was printed from many big cities like Paris, La Haye, Leipzig, Copenhagen, etc. In the volume, there were eighteen separate maps on India.

Jean Nicolas Bellin (1703-72) was a serious and scientific cartographer and was attached to the French Marine Office. He generally used well-founded information only as far as practicable. A large map of the peninsula, *Carte reduite de la Presque Isle de l'Inde*, showing mostly places where navigational facilities were available, along with other marine maps, was published in the *Hydrographie Francaise* in 1756. Various editions came out thereafter till 1776. During 1763 and 1764, a large atlas—*Petit Atlas Maritime* containing 581 maps, including 18 of the Indian coast, was published. Five of his finest maps have been included in this volume. These were: (1) *Carte reduite de la Presque Isle de l'Inde*, (2) *Carte de l'Indoustan*, (3) *Coste de Malabar, Canara*, (4) *Suite de la carte de l'Indoustan*, and (5) *Carte des Costes de Concan et Decan*.[43]

The map No. 33A in the volume shows northern India including part of Afghanistan. The portion extending from Central India up to the Orissa coast is almost blank. All these maps were compiled with Paris as the prime meridian.

In 1763, Robert Orme, a personal friend of Robert Clive, Governor of Bengal, published his *History of the Military Transactions of the British Nation in Indostan*. For the first volume, he used maps of India engraved by Thomas Kitchin. Robert Orme had first-hand knowledge of India and Indian affairs as he was employed in India by a mercantile firm for about 15 years from 1743 to 1758. He had the privilege to stay at Calcutta and Madras. Clive

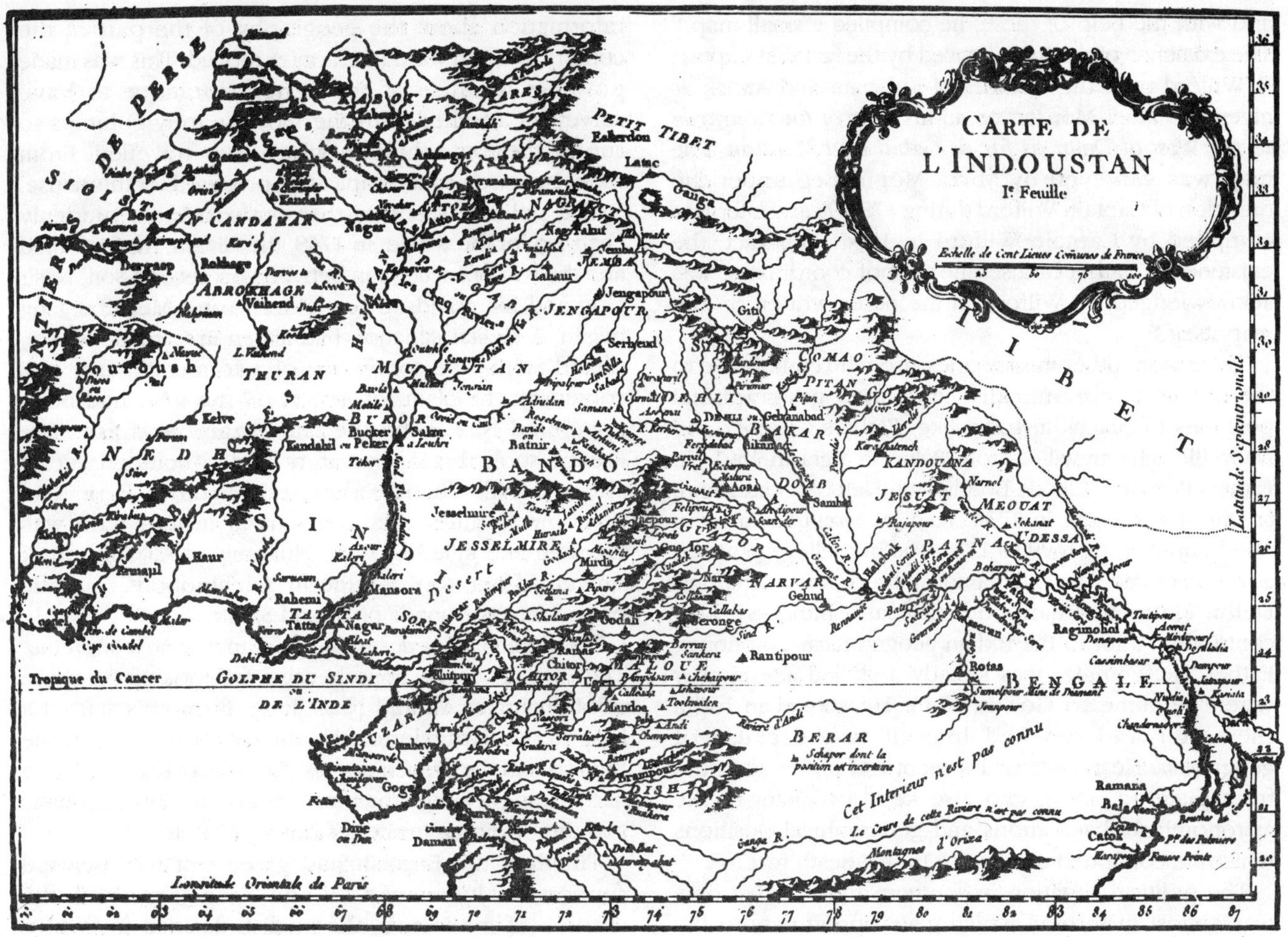

Map 18: Detail of *Carte de l'Indoustan* by Jean Nicolas Bellin. (Courtesy: Susan Gole, *A Series of Early Printed Maps of India in Facsimile*, New Delhi, 1984).

and his successor Vansittart helped him to get official maps for his personal use, which were compiled at state expense.

These were some of the well-known map-cum-travel accounts published during the eighteenth century.

Though the political equation among the European nations during the war of Spanish Succession, was disturbed in the European continent from 1702 to 1713, it left India untouched. There was sufficient amity being displayed by two estranged nations, the English at Fort St. David and the French at Pondicherry, during the war between the English and the Raja of Jinji in 1712. But this political balance was disturbed in the Indian subcontinent during the later half of the eighteenth century. This was mainly due to the political situation which tempted the foreigners to meddle in the political affairs of the native rulers and forced them, particularly the English and the French to intervene on opposite sides.

Till the end of the eighteenth century, route surveys were the only basis for a good topographical map of a territory. The route surveys, wherein the geographical positions of most of the places were astronomically fixed and the distances between them accurately measured or calculated from the meticulously measured points, with their compass bearings, were the ideal requisites to get to know the topography of the place.

The value of the astronomically calculated bearings indicating the geographical positions of places had been recognised and accepted as early as the second century AD. Ptolemy was, perhaps, the first to use them systematically. He had left a long list of places with their geographical coordinates. In later centuries, many other geographers had compiled tables of such coordinates.

Ptolemy's list of coordinates also included some coordinates of places in India, which was supplemented later by other geographers.

In the more recent times, we know Father Monserrate travelled from Surat to Fatehpur Sikri in 1580 and in 1581, from Fatehpur Sikri to Jalalabad, in Afghanistan, in the company of the Mughal Emperor Akbar. He compiled a list of over 100 such geographical coordinates of places and with the help of these, he compiled a small map.[44] The existence of the list is proved by the fact that Captain F. Wilford used the latitudes of Kalanaur and Attock as given by Father Monserrate, in his *Map of the Countries to the West of Delhi as far as Cabul and Moultan.* The map was surveyed by Mirza Mogul Beg under the direction of Captain Wilford during 1786-96 and later was compiled by Captain Wilford in 1804. This fact, the existence of Father Monserrate's list of coordinates, was acknowledged by Wilford in the notes written on the map itself.[45]

There were other missionaries also who contributed to the list of astronomically calculated geographical positions of places in India like, Fathers Grueber and D'Orville who travelled from Pekin to Agra in 1661-62, Fathers Pons and Claude Bondier who left Chandranagore to Jaipur and back in 1734. Raja Jai Singh Sawai who ruled Jaipur from 1699 to 1734 was himself a renowned astronomer and who built famous observatories at Jaipur, Delhi, Mathura, Ujjain and Varanasi (Banaras), also contributed a lot to the Indian geographical astronomy. Father Tieffenthaler was already a skilled astronomer when he came to Goa in 1743. He stayed in India thereafter and devoted himself to the cause of geography. He maintained a record of all his journeys from place to place, and also kept a register of the astronomical observations and geographical positions which he continued to observe till his death in 1785.

The political situation in southern India at that time was almost in turmoil which gave opportunity to the English and the French to interfere to gain economic advantages out of it.

Towards the close of the first quarter of the eighteenth century the Mughal Emperor had lost all control over southern India. Asaf Jah, the Nizam-ul-Mulk, was the Subedar of the Deccan, who had practically made himself independent of Delhi by 1724. The Carnatic was a Province under the Subedar of the Deccan, and was ruled by a Nawab with his capital at Arcot. He was also practically an independent ruler. The political affairs of the Carnatic presented a scene of disorder and confusion.

At this time, the War of Austrian Succession broke out in Europe in 1740. The English and the French, in Europe were in the opposite camps. An echo of this situation was heard in India as well. Troubles began in 1746 when an English fleet appeared on the Coromandel coast and threatened Pondicherry, the French settlement. The French appealed to the Nawab of the Carnatic, Anwar-ud-din, who forced the English to withdraw. After some time, a French fleet attacked and captured Madras, the English factory. In 1748, war was brought to a close by a treaty and Madras was restored to the English. This was the first engagement between the English and the French, on the Indian soil and it gave the respective armies an opportunity to survey and have first-hand direct information about the geography of the part of the country through which the armies moved. This was made possible because of the general practice to have surveyors attached with the moving army columns to survey the route through which they travelled. From these exercises, route maps were prepared for future use.

The political scene in the southern India suddenly changed for the worse. In 1748, the Nizam, Asaf Jah died and the war of succession between his second son, Nasir Jang and his grandson (daughter's son), Muzaffar Jang began. A similar situation had arisen in Carnatic where, Chand Sahib, the son-in-law of a former Nawab, was contesting the claim of Anwar-ud-din who had been appointed Nawab by the late Nizam, Asaf Jah. The French took this opportunity and concluded secret alliance with Chand Sahib and Muzaffar Jang. The combined armies attacked Anwar-ud-din who was defeated and killed in 1749. Muhammad Ali, the son of Anwar-ud-din, took refuge in Trichinopoly. Carnatic passed into the hands of Chand Sahib.

The French influence in the Deccan was now supreme. The English became apprehensive and took up the cause of Muhammad Ali and Nasir Jang. From this started a long-drawn struggle for the throne during which the English and the French armies clashed several times and had to trample under their feet, almost the entire Carnatic and the adjoining territories during 1746 to 1763.

Till then the English had given but half-hearted support to Muhammad Ali and Nasir Jang. With the victories of the French, the English decided to involve themselves fully into the wars. Intrigues and counter-intrigues went on. Many battles were fought; claimants were murdered and changed sides and the ultimate result was that Chand Sahib, a French supported ruler, became the Nawab of Arcot. Muzaffar Jang was installed as the Nizam; but he was killed shortly after and Salabat Jang, another French nominee, became the Nizam. Clive came on the scene; he planned the relief of Trichinopoly, attacked Arcot, forced Chand Sahib to lift the Trichinopoly siege. Marathas were also brought into the fray. Clive's strategy turned the tables upon the French who, along with their allies, were completely defeated. Chand Sahib surrendered to the Raja of Tanjore and was later murdered. Muhammad Ali became the undisputed Nawab of the Carnatic. As stated above, the English armies had to move about the whole of the Carnatic which afforded them the opportunity to traverse almost the entire country.

Since both the warring factions in India were exhausted, financially and otherwise, the Directors of both the companies held several meetings and long-drawn negotiations for peace in India which was ultimately achieved, though provisionally, in January 1755. Ratification of this peace from the respective headquarters in Europe, never came. In the meanwhile,

Seven Years War in Europe came about in 1756. Since the English and the French in India were already exhausted, there was forced peace in India.

During this interval, trouble broke out for the English in Bengal. Siraj-ud-daula attacked and captured Calcutta. Clive sailed from the Coromandel coast for Bengal to restore the English authority. The Northern Circars was attacked and the French forces were defeated at Condore in December 1758. In early 1759, Masulipatam was captured. Salabat Jang, ruler of Northern Circars, ceded Masulipatam along with eighty miles long and twenty miles wide territory to the English. The enmity between the two European powers continued. The French finally surrendered in January 1761. By the Peace of Paris, in Europe, in February 1763, Pondicherry was restored back to the French, and with this, their influence on the Coromandel Coast and in Bengal, was considerably, reduced. The Northern Circars had passed under the authority of the English in 1765. Thus, in southern India, the Anglo-French conflict lasted for about twenty years.

As has been noted, there were many geographers who had prepared quite accurate maps of the routes they had travelled by observing geographical coordinates and measured or estimated distances. The earliest record of such activity by a servant of the East India Company has been recorded by James Rennell in his *Memoir of a Map of Hindoostan.* This was a traverse survey of a route from Devicote, a fort at the mouth of Coleroon river, captured by the English in 1748 which, ultimately, was granted to them by the Raja of Tanjore. This town has since been washed away. The route was from Devicote to Trichinopoly, via Tanjore. The survey was carried out sometime before 1651 by John Barker.[46] This was the beginning of hundreds of such route surveys that were carried out by surveyors attached to the armies of the European nations, more especially the English, that criss-crossed the country in their numerous battles which they had to fight before they could subjugate India. There is mention of a French route map from Palamcottah through the *Royaume de Madura* to Trichinopoly, on a scale of 3 miles to an inch. Though undated, it appears to belong to the period between 1753 and 1754 when several of such maps were drawn by a French national, George Erbb, then serving in the British army of Major stringer Lawrence, stationed at Fort St. David. Lawrence was very active in the skirmishes with the French around Trichinopoly at that time.[47] Rennell had used many such surveys while compiling his *Map of Hindoostan* and other maps. He has given details of such route traverses in his *Memoir of a Map of Hindoostan*.[48]

During this long war which went on almost incessantly from 1746 to 1763, with two short intervals of about three years each, the English came out ultimately victorious. Their influence prevailed from the north of Northern Circars to Cape Comorin in the south, covering the region of Arcot, Tanjore and Northern Circars in the peninsular India. This war gave them an opportunity to collect geographical information of the districts of Trichinopoly, Tinnevelly and Madura by military route traverses under the direction of the Chief Engineer, John Call. With the help of the materials thus collected, a general map of the Southern Peninsula was compiled by Henry Montresor[49] in 1770, which Rennell and other surveyors acknowledged as an authority for many years.

In 1763 the Nawab of Arcot, also called the Nawab of Carnatic, Muhammad Ali, ceded to the English the district of Chingleput which became generally known as the *Jagir* of the English.

On hearing of this grant, the Directors at once ordered the district to be surveyed in detail. The surveyor was Thomas Barnard who was ordered to survey the territory without the help of an assistant. He was expected to survey the area topographically and include therein all the information required for revenue purposes as well. There were many hindrances and the officials of the Nawab did not extend as much cooperation as was expected of them.[50] The survey, along with map and the reports,[51] could be completed only by November 1774. The map was originally laid down on the scale of 2 inches to a mile, but was later reduced to the quarter-inch scale. This reduction was sent to the Directors and was published in London by A. Dalrymple in 1778.[52]

The survey of this land, about 100 by 50 miles, was really a creditable job done by just a lad of 19 years, single handed. To overcome certain difficulties and to fulfil the requirements of a full and detailed survey, he carried close circuit traverses which were carried right to the bases of triangulated hills, where these were available, with least closing errors.

By this time the English East India Company had entangled itself deeper and deeper into the political and military affairs of the local Indian rulers. So, surveys were continually required of the regions surrounding the countries under their influence.

In 1761, Haider Ali became the ruler of Mysore and was expanding his dominions and influence almost in every direction. Consequently, there was enough trouble all along the frontiers of Carnatic and Mysore. Carnatic, being under the protection of the English, the Company was naturally anxious. They were more worried because the boundaries of the Nawab and the Mysore were not clear and were not defined, resulting in constant disputes. Officers were, therefore, ordered to find out the details of the boundaries of Karur, Manapar and Trichinopoly districts. Sketches of the areas and the passes leading into the Carnatic, were prepared and the region explored. The survey was undertaken, obviously, from the Carnatic side only. James Rennell praised the efforts of Captain Bonjour who prepared the map entitled *An Account of the Passes between the Parallels of*

Udeghery and Sautgud, in his memoir.[53] Some maps of the region lying on the west of the Carnatic, and Seringapatam were prepared as a result of first Mysore war of 1767-9. Since 1770, Robert Kelly had started surveying all the routes in the Carnatic which he had the occasion to march through.[54]

Keeping in view the deep hostility of Haider Ali of Mysore, and the constantly disturbed state of the region around Carnatic, the Company persistently requested permission from the Nawab of Carnatic to let them survey the entire country, but was refused.

Between Rajamundry and Vishakhapatnam, the routes were surveyed and a large map covering Colonel Forde's marches was prepared. There was another survey made between Vishakhapatnam and Coopilly; Major Polier prepared a very accurate map of the route between Vishakhapatnam and Jagarnaut (Jagannath) Pagoda at Puri in 1759 on his return journey to Bengal from the Masulipatam expedition.[55]

In southern India, in 1765, Emperor Shah Alam granted the *dewani* of the Northern Circars comprising Kondapalli, Ellore, Rajamahendri and Chicacole to the English; and in 1766 a treaty with the Nizam authorized the English to occupy all of them except Guntur Circar which was the *jagir* of Basalat Jang at that time and was ultimately ceded to the Company in 1788. The occupation of the ceded Circars was delayed because war broke out with Haider Ali, who had earlier bought off the Marathas. The Nizam too was won over by him. Thus, both the partners of the English were detached away from them. Haider Ali and the Nizam both were defeated at Trincomalai Changama pass in 1767. In the meanwhile, the English forces from Bengal invaded Northern Circars, the Nizam's territory, reached Ellore and then Warangal, and were poised to attack Hyderabad. Treaty with the Nizam was then signed at Masulipatam in 1768. Ganjam was later occupied by the English. War with Haider Ali however continued. He ravaged Carnatic, defeated Hector Munro who retreated to Madras and by a surprise march following Munro reached within a few miles of Madras. The Madras Government in their fright, concluded peace with Haider Ali in 1769. This is known as the first Mysore war.

W. Stevens had earlier prepared many maps of a large portions of the Southern Circars. These maps were well received later by Rennell and others. He was again asked, in 1773, to make a survey of the entire frontier of the Southern Circars facing the Nizam's territory. Simultaneously, Cotsford was ordered to survey Ganjam.[56]

Apart from the maps prepared from the traverses conducted during the wars, mostly in the Carnatic and the regions immediately on its western borders, the English army was constantly demanding more and more detailed maps of the Carnatic.

Robert Kelly was busy with his plans of route surveys as and when he was able to spare time for them.

In March 1777, General Joseph Smith, the Commander-in-Chief of the East India Company, submitted to the Government a collection of road surveys and maps in the Carnatic, traversed by John Pringle and bound in the form of a book called the *Book of Roads*. He recommended that Pringle be employed to survey some other roads in the Carnatic. He further stated that John Pringle was practically the 'Master of the Roads and Situations' in the region.[57] Two months later, another book of roads, prepared by Pringle, was sent to the Government. The Commander-in-Chief recommended him for the post of Captain of Guides and Hircarrahs. He was appointed Captain of Guides to the Army in 1777 and Captain of the Sepoy Corps in 1780. The Corps of Sepoy Guides was now a regular establishment. It did valuable work of gathering geographical information for the next about 25 years. In 1780, Pringle sent in to the Government his third book of roads.

In the war of July 1780 with Haider Ali, John Pringle surveyed the marches of the army under Sir Eyre Coote and considerably extended the surveyed portion of the southern peninsula. James Rennell described it thus: 'By means also of Mr. Pringle's bearings, and measured routes the positions of Porto Novo, Sautgud, and Amboor[58] are obtained; which last may be considered the most westerly point determined with accuracy, anywhere north of Trichinopoly.'[59] After the war, John Pringle compiled yet another book of roads of the Carnatic which he himself measured. One copy of the map that accompanied the book is entitled *A Chart of the route of the Grand Army in the Carnatic under the Commands of Major General Sir Hector Munro, K.B., Lieut. General Sir Eyre Coote K.B., and Major General James Stuart from August 25th, 1780 to June 7th, 1783 ... John Pringle, Captain of Guides to the Army on the Coast of Coromandel. Copied and drawn by J. Reichel 1787, Scale 1 inch=2$\frac{1}{2}$ miles.*[60]

There was yet another book of *Roads in the Carnatic, ... 1786-87.*[61] he had surveyed, apart from the road traverse from Madras to Masulipatam. The topography of the country adjacent to the roads, was shown in some detail. The book also showed the villages, courses of rivers, tanks, hills and passes, forts, topes and jungles of every kind, the best places for encampments, along with remarks that could be of some use to the army officers such as the availability, in sufficient quantity, of fresh drinking water, grains in the villages, etc. He had also mentioned the geographical coordinates of many places including those of the mouths of rivers, etc.

We have seen that during the Anglo-French conflict in the Carnatic and the post-war period, almost all the major routes in the area, and those leading out of it to the north, the south and the west, along with the passes leading therefrom covering Northern Circars, Carnatic or Arcot and Tanjore, were surveyed by the army surveyors. Their efforts in this regard were appreciated by the future Surveyor-General of Bengal, James Rennell, as these were very helpful to him in the compilation of his maps of the peninsula of India.

In southern India, an era of major troubles and bloody wars began. The three main native powers, the Nizam, Mysore (after 1761) and the Marathas were almost constantly at intrigues and military confrontations with the English, involving them, sometimes singly and at other times changing sides and even fighting jointly as partners.

It all began with Haider Ali becoming the ruler of the Mysore in 1761. Troubles in southern India, erupted one after the other in almost every corner of the peninsula. The result was that the armies of the English, the French, Mysore, the Marathas and to a lesser degree of the Nizam, were almost constantly at march against one or the other.

With the crumbling to pieces of the once mighty Mughal Empire, most of the native rulers began asserting their independence, not only in the east but throughout the peninsula as well.

Asaf Jah, the Nizam-ul-Mulk, who, on behalf of the Mughal Empire, controlled the rulers of the south died in 1748. By this time imperial power had also weakened. The initial result of this was the Anglo-French conflict, mainly over the Carnatic, and the ultimate rise of the English military might, which made the English, a territorial power as well. Although their territory was very small, it was significant. The English opinion was not to be ignored in future conflicts.

Now the Marathas, under the vigorous administration of the Peshwas, had asserted themselves and had extended their authority over a vast territory, from the southern banks of the Chambal river in the north-west, the Ganges in the north, bordering Bihar and Bengal on the north-east, touching the Bay of Bengal in Orissa, to Concan on the banks of the Tungabhadra river, bordering Nizam's territory on the left banks of the Godavari river. There were other influential Maratha chiefs, like Raguji Bhonsle, Anand Rao Pawar of Dhar, Damaji Gaekwar, Mulhar Rao Holkar and Ranoji Sindhia, who could not be ignored. Apart from their sovereignty over their territories, they had imperial authority to collect *Chauth,* i.e. one-fourth of the revenue from a number of rulers, like Gujarat, the Deccan in the south and provinces of Lucknow, Patna, Bengal, Allahabad, Agra and Ajmer. Fortunately for the English, these Maratha chiefs were not united under one ruler; their *Raja,* Ram Raja, a titular King, resided at Satara. The real power was taken ever by his Prime Minister or Peshwa, as he was called, who had his headquarters at Poona. There was at that time, what may be termed as a loose 'Confederacy' of the Maratha chiefs.

We have noticed that during the Anglo-French conflict in the Carnatic, only one of the three Presidencies, namely Madras, was mostly involved in the wars, whereas the Bombay and the Bengal remained comparatively less embroiled except a couple of times, when a relatively small force was sent from Bengal by Clive to Madras' help, which invaded Northern Circars. But now, during the conflicts with the new alignments of forces that had emerged, armies of all the three Presidencies, Bombay, Madras and Bengal, were involved. Even the Governor-Generals from Bengal took personal command of forces a couple of times during the eighteenth century to crush their enemies in peninsular wars.

All the conflicts with these powers in the southern peninsula, during this century, were interconnected. To examine the movements of the various armies and the *consequent progress of route traverses in the region,* it would be proper to know, in some detail, the wars with those powers individually, as far as possible. These route traverses were the major sources of obtaining technical data required to prepare scientific and accurate maps of the areas as per the technology in vogue.

We have already referred to the first Mysore war (1767-9). The area which the main force of the English army had marched through, was the Carnatic and Baramahal.

Three years after the conclusion of the treaty with Haider Ali in 1769 at Madras, war with the Marathas, also known as the first Maratha war (1772-82), broke out.

This war was the result of the dispute for succession to Peshwaship at Poona by two rival claimants, Nana Fadnavis and Raghoba. We may not go into the details of the intrigues here, but in brief, Raghoba applied for English help. The Bombay Council readily agreed and the Treaty of Surat was signed in 1775. But Bengal Council, the supreme body, did not approve of it and a fresh treaty was signed, the Treaty of Purandhar in 1776. The Directors in England, however, approved the earlier Treaty of Surat. The Governor-General Warren Hastings, on the other hand, kept up his alliance and went on favouring Raghoba.

In 1779, an English force marched from Bombay to Poona. This was defeated by the Marathas at Telegaon. The convention of Wadgaon, was signed in 1779 by the Bombay army. This was however refuted by the Governor-General, Warren Hastings, and the war continued. A strong force, under Colonel Goddard, was sent against the Marathas. Colonel Goddard marched from Kalpi on the river Jumna, to Surat through Central India. On his way, he captured Ahmadnagar and Bessein and then advanced towards Poona. A part of the English force, captured Gwalior. Sindhia was again defeated at Sipri. Peace was established with Sindhia. At this time, another war was simultaneously being waged by the English against Haider Ali. On Sindhia's mediation Treaty of Salbai was signed with the Marathas at Poona in 1782.

Three years before the Treaty of Salbai, Haider Ali, who was looking for an opportunity to take revenge against the English because they had refused to help him, in spite of the treaty obligations of 1772, against the Marathas, got the opportunity in 1779. He joined in coalition formed by the Nizam in 1779 against the English, and the war broke out. Haider Ali burst upon the Carnatic with fury, forced Colonel Baillie to surrender; Munro retreated in panic to Madras from Conjeveram; Arcot, the capital of Carnatic, was lost by the English. In the meanwhile, Hastings won over the Nizam; the treaty with Sindhia at Salbai in May 1782 was signed; Sir Eyre Coote was sent by sea from Bengal who defeated Haider Ali at Porto Novo on 1, July 1781. A strong force under Colonel Pearse also marched down the east coast from Midnapore in Bengal on 21, January 1781. The Dutch had joined hands with Haider ali because of their hostility to the English in Europe. Haider Ali defeated General James Stuart and took Cuddalore from the English and his son, Tipu, gave a crushing defeat to Colonel Braithwaite in 1782 in Tanjore. In the meanwhile Haider Ali died. Tipu continued the war. In 1783, the hostilities in Europe between England and France came to an end, and consequently they came to terms in India as well. Tipu fought alone, and captured Bednore and Mangalore. In November 1783, the English captured Palghat and Coimbatore. The Treaty of Mangalore was signed in March 1784. Thus ended the second Mysore war.

Several significant additions were made to the geography of India during this war. T.D. Pearse, himself being a surveyor, and a keen and experienced astronomer, '... was engaged in making astronomical observations at Calcutta from 1774 to 1782, and in the latter year he undertook a journey to Madras, for the purpose of fixing positions, and laying down the intermediate coastline for a distance of 700 miles. He observed for latitudes by stars' meridian altitudes, for longitudes by immersion of Jupiter's satellites, and measured the distance with a perambulator. This work occupied two years, from 1782 to 1784, ... ably assisted by R.H. Colebrooke, the future Surveyor-General who latterly took all the observations....'[62]

But, another eminent author[63] has stated that the main observations and perambulator traverse were taken by Colebrooke on the return of Pearse's force from Madras in 1784.

Under his personal direction, Colebrooke ran a continuous traverse with the help of a perambulator, noting down distances travelled from one stage to another. On the way, he regularly took a series of astronomical observations for latitude and longitude.

Further, Robert Kelly and other surveyors on their march in 1783, with Colonel Fullarton's army from Negapatam on the east coast, surveyed a route through Madura to Palghat, the place which was reached only the previous year, by surveyors under Colonel Humberstone's detachment, thus connecting the two important lines of surveys.

Finally, Nagpur was connected by surveys of James Ewart who was attached with the mission sent by the Governor-General to the Raja of Berar to secure his friendship against Mysore.

Two more wars in the southern peninsula remained to be waged before the close of the eighteenth century, third and fourth Mysore wars. The third Mysore war was fought during 1788 to 1792 and the fourth and final war with Mysore in 1799. The scene of operations, obviously, was southern India.

In 1788, the English obtained Guntur from the Nizam. Earlier in 1784 by the treaty of Mangalore, they had conceded the right of Tipu Sultan over the territory which the Nizam now claimed and asked the English to help him to recover it from Tipu. The English tried to confuse the issue. Tipu got enraged and attacked Travancore, an ally of the English. The Governor-General himself took command in December, 1790; captured Bangalore in March 1791; defeated Tipu outside Seringapatam in May and then retreated. A second advance towards Seringapatam was made in January 1792; outskirts of the town were captured and a treaty was signed in March 1792. Tipu was forced to cede nearly half of his dominions. The English obtained Malabar and sovereignty over Coorg, on the west of Mysore. On the south, Dindigul with its surrounding districts, and on the east the Baramahal Passes overlooking Carnatic were acquired. The Marathas gained territory to the north-west and the Nizam on the north-east of Mysore. This was the result of the third Mysore war.

Ever since the treaty of Seringapatam of 1792, Tipu was intriguing against the English and negotiating with the French. Ultimately, Wellesley, the Governor-General, sent two armies against Mysore; the main army under General Harris, by way of the Carnatic; the other the Nizam's contingent from Hyderabad, under Arthur Wellesley, where it was stationed as part of the Subsidiary treaty that had taken place in 1 September 1798 between the Nizam and the English. The second force attacked from the east and the Bombay army made its way from the Western Ghats. Tipu was first defeated by the Bombay army and then routed by General Harris at Malavelli. Tipu retired to his capital, Seringapatam, which was stormed on 4 May 1799. Tipu was killed. This was the last Mysore war.

Most of the territories were annexed. Kanara on the west coast; Coimbatore on the south and some districts on the east, together with the fort of Seringapatam were brought direct under the English control. The Company thus had continuous territory in the south, from the

Malabar coast on the west, to the Coromandel coast in the east. Thus the Mysore power was completely crushed from Indian politics.

The Marathas still remained quite an influential force in India and had in fact spread to the centre and northern India as well. However this will be discussed in the next chapter.

We have seen that it was in the Maratha war of 1772-82, that the Bombay Presidency was also involved. Though its role in giving military assistance to Madras was a small one, geographically, it added considerable new information to the already known geographical facts.

Even after the treaty of Salbai in 1782, hostilities between the English and the Marathas continued. General Mathew from Bombay captured Mangalore and Onore (Honavar), both on the coast of Kanara, a part of Mysore, and marched towards Bednur which commanded the approaches into Mysore over the Ghats. Charles Reynolds was attached with the army as surveyor. The English were defeated, and Reynolds escaped to Onore which was still under the English. After peace was signed in March 1784, he reached Bombay. During the time he was at Onore, he compiled his well-known map of Bednur which was greatly appreciated by James Rennell,[64] the Father of Indian Surveys; and the first Surveyor-General of Bengal, 1767-77.

In 1785, a political mission was sent to meet Mahadji Sindhia who was then with his army in Central India. The English mission, starting from Surat, passed through Ujjain, Gwalior, through the Plateau of Malwa, and found the Maratha chief near Mathura. The Mission then visited Delhi and marched to Kanpur; from there, they travelled down the river to Calcutta. This line of survey was greatly needed at that time. The Mission ultimately reached Bombay by sea.

Charles Reynolds was asked to accompany C.W. Malet, who was being posted as the English Resident at the Court of Peshwa at Poona. A special request was made by the Governor-General to him to carry out his work of survey in such a manner that it might not offend the Marathas.

During the period, he travelled much in the Deccan for traversing major routes. He always followed different roads while going up and down his journeys between Poona and Surat during 1788-9. He travelled from Nagpur to Hyderabad; from Masulipatam to Madras, returning again via Hyderabad just to enable him to complete the survey of the region. In 1789, because of the impending hostilities with Mysore, he was asked not to move out of Surat, where he began compiling his maps.

During the third Mysore war described earlier, James Allan and Colin Mackenzie surveyed new routes along the southern boundary of Mysore.

Before moving to Madras to take command of the forces, the Governor-General sent the Surveyor-General, Alexander Kyd to Madras to gather whatever information he could regarding the approaches into Mysore. The Corps of Guides were already busy collecting such information. In January 1791, Colebrooke, who was appointed surveyor to the Grand Army, carried a continuous line of survey from Madras, through Vellore right into the Mysore territory. For this purpose, he ran a series of triangles from the beginning to the end, i.e. from Madras to Seringapatam; took several astronomical observations and perambulator measurements of the route.[65] Although his maps were very neatly drawn, they were not complete, as they depicted routes and areas close to them and distant topographical features visible from the surveyed routes only.[66]

The Governor-General, during the war, had secured the cooperation of the Nizam and the Marathas. To bolster up the Maratha army which was marching south from Poona, a detachment of the Bombay army, under Captain Little, joined the main army in May 1790. On reaching Dharwar, their advance was held up. On November 25, 1790, reinforcements from Bombay landed at Cardona on the Jaigarh river under the command of Colonel Charles Frederick, with G.L. Emmitt as the surveyor. Emmitt at once took astronomical observations at the mouth of Jaigarh river and ran a traverse line to Dharwar. After the fall of Dharwar, Emmitt was attached with Captain Little's detachment and Colonel Frederick accompanied the detachment to Mysore. The rest of the army marched back to Bombay via Poona.[67] The Maratha army joined the Grand Army north of Seringapatam. G.L. Emmitt, who was constantly engaged in survey work, made junction with the traverse line brought by Colebrooke from Madras, and then continued his line to Bangalore, Sira, Harihar, then to the junction of the Kistna and Tungabhadra rivers, again to Pangal in the Nizam's territory and through Badami, Dharwar, to Goa. This line of survey was considered to be of the important traverse surveys linking east coast with the west coast and moving east-west through the centre of the peninsular India. With this, Emmitt also became the first English surveyor to reach Goa by land.[68]

After the end of the war, the surveyors were deputed to survey every part of the territory that the armies had marched through, or that was left unsurveyed or was not properly traversed.

Alexander Kyd traversed a line from Seringapatam towards the west coast of India, over the Ghats and Coorg, to the coast and closed it at Anjengo via Cannanore.

Anbury and James Blunt surveyed a line to Kalpi, surveying through Hyderabad, Berar and Saugor.

Charles Reynolds was asked to survey another

important route from Hyderabad to Agra.

After the close of the war in 1792, Malabar was given to Bombay Presidency. G.L. Emmitt was at once asked to start its survey. John Mather surveyed Salem district including Baramahal. Incidentally, this was the first district survey which was based on a system of theodolite bearings and intersections as opposed to the usual perambulator traverses. It took almost five years to complete.

Colin Mackenzie was appointed in 1792 as surveyor with instructions to devote himself to the geography of the Deccan.

Perhaps, one of the important surveys of this period was the traverse line surveyed by James Blunt from Chunar down through the Central India, through the headwaters of the two important rivers of the region, Son and Mahanadi, down the rivers Wainganga, Wardha and Godavari to the coast of the Bay of Bengal.

For giving a final blow to Tipu Sultan, the Governor-General once again moved from Calcutta to Madras in 1799. He did not take to the field himself but took personal control of the war. Alexander Beatson was made the Honorary Surveyor-General to the Grand Army. I have already given the details of the exploits of Bombay Army and the other under Arthur Wellesley, the Subsidiary Force with the Nizam. Seringapatam was captured and the fort was stormed. Tipu was killed in the action.

Mysore was shorn of the outlying districts. Mackenzie was ordered to immediately start the survey of Mysore and the newly ceded districts by an order of the Governor-General issued in September 1799, even before he returned to Calcutta.

By the end of the fourth Mysore war, the power of Mysore had been crushed for ever; and the districts of Kanara, Coimbatore, Waynaad and the Nilgiri Hills were annexed by the English. The Nizam was weakened by the cession of districts of Bellari, and Cuddapah along with the territories to the south of the Tungabhadra river and to the south of the Kistna below its junction with the former. Surveys of these newly acquired territories were immediately taken up. The power of the Maratha Confederacy was, though not crushed was considerably weakened as the various Maratha chiefs and the Peshwa could not unite among themselves. This was the political scenario in south India after the end of the fourth Mysore war in 1799.

We have also observed how extensively the route surveys were taken by traversing thousands of miles, covering hundreds of offshoots of the main lines. The geographical material now available was so scattered and varied in nature that serious discrepancies were inevitable—survey lines winding through hills and jungles, the distances, obviously, roughly measured, and sometimes only estimated, latitude observations, of course of valuable help where available, and the longitude observations very much uncertain. But huge material for the geography of the peninsula of India was now there for the geographers and cartographers to exploit.

It is this material, which James Rennell later used very skilfully in his maps, that won him great respect as father of Indian Geography.

In the seventeenth and eighteenth centuries military activity in the northern India was not as extensive as in the southern peninsula. But the region was surveyed in comparatively peaceful atmosphere, and, therefore, in a more scientific way. This was due to the fact that the period between 1756 and 1761 saw the transformation of the East India Company, from a trading concern, into a territorial power in India.

Military movements in the northern region began with the battle of Plassey in 1757. Earlier, topographical knowledge was confined mainly to the banks of the rivers, notably, Ganges and Hughli. As a result of the victory in Plassey, the newly installed Nawab of Bengal, Mir Jafar, ceded to the English the whole of the territory of 24-Parganas under a formal treaty of 20 December 1757. This acquisition of territory in Bengal was bigger than the lands in Calcutta acquired earlier. Uptil now, they were regarded as a body of prosperous merchants with one rich settlement, a few territorial rights in some villages round Calcutta, and some agencies or factories at Cossimbazar, Dacca, Balasore, Jagdea and Patna. Without going into much detail, it may be stated that Mir Jafar, the Nawab of Bengal, was forced to abdicate in favour of his son-in-law, Mir Kasim. On the promise of military aid to him, the new Nawab, in return, further ceded to the English Company, by a treaty of 1760, the districts of Chittagong, Burdwan and Midnapore.[69] Because of the war of succession in Bengal, the political situation in the northern India at that time, was fluid. At one stage, in 1764, Shuja-ud-daula, the Nawab-Vazir of Oudh, combined with the Mughal Emperor, Shah Alam, and Mir Kasim, the Nawab of Bengal, all combined together against the English, and the result was the war of Buxar, in 1764. The combination was defeated by the English; Kora and Allahabad districts of Oudh were acquired by them. Mir Jafar was again installed as Nawab of Bengal. After the battle of Buxar, a defensive alliance was concluded with the Nawab of Oudh. The idea behind this was to make Oudh a buffer state against the Marathas, who had reached the Jumna river by then and were posing a threat to the English possessions in the north. Shah Alam, formally made over the districts of Kora and Allahabad to the English and the rights to collect *Diwani*, i.e. to administer and collect the revenues of the provinces of Bengal, Bihar and Orissa were also given to the Company.

In the meanwhile, the Maratha power felt a jolt. They were defeated at Panipat in 1761, by the Afghan ruler, Ahmad Shah Abdali. Shah Alam who was under the protection of the English left them and joined the Marathas. In the meantime, Robert Clive had left for England in January 1767. It was during the administration of Warren Hastings (1772-85), that the Marathas had recovered their power and were again making themselves prominent both in the north and the south. In the south, Haider Ali was also thinking of revenge against the English. Warren Hasting's priority, like that of his predecessor, Clive, was to safeguard Bengal and so he also tried to make Oudh a strong buffer state. He concluded a treaty with Shuja-ud-daula, the Nawab-Wazir of Oudh; sold Kora and Allahabad to him; and was promised English support, in case of war.

Rohilkhand was also under a threat from the Marathas and the Rohilla chiefs also concluded a treaty in 1772 with the ruler of Oudh. In 1773, Rohilkhand was invaded by the Marathas. Oudh forces along with the English, because of the treaty obligation, came to the help of Rohilkhand. Marathas, seeing the combined forces, retreated without a war. Oudh demanded compensation from Rohilkhand as per the treaty obligations, which the Rohilla chiefs evaded. So Oudh asked for English help to recover payment. Troops under Colonel Champion were sent and the Rohillas were defeated at Miranpura Katra in 1773 and Rohilkhand was annexed to Oudh.

Lord Wellesley (1798-1805) forced upon the Nawab of Oudh a new treaty in 1801 by which the Nawab ceded Gorakhpur, Rohilkhand and the territories lying between the Ganges and the Jumna, known as the Lower Doab to the English. This was the story, in brief of the military campaigns in the north during the seventeenth and eighteenth centuries.

Another source of gathering geographical information was the routes followed by political missions sent to the courts of Indian rulers. Such missions too had surveyors with them to survey the roads traversed. Such occasions were always so arranged and their routes and the time of their despatch planned in such a way that it enabled them to travel through hitherto unknown countries, obviously to gather the maximum geographical information about new and previously unexplored regions.

It was in 1775 that Rev. William Smith was asked to accompany Colonel Upton's political mission to Poona. Smith was an astute astronomical observer. Practically, every night he used to note down the observations made during his survey across the unknown regions of India. This embassy to Poona afforded him an excellent opportunity to survey most of the western part of the peninsula.[70] The mission started from Kalpi on the Jumna river on 24 October 1775. The route taken was the well-known route normally being followed by travellers and traders while going to the South during that century. The party travelled to Narwar, then to Sironj, crossed the Narbada river and reached Burhanpur, the capital of Khandesh; again, crossed the river Tapti to Ahmadnagar, and finally reached Poona. While at Poona, Smith got an opportunity to go to Bombay. With the help of the material so collected along with astronomically calculated coordinates of many places, he began constructing a map which could not be completed by him while he was in India, due to ill health. So he took the material to England. James Rennell has spoken very highly of this map. It was, according to him, for the first time that a dependable geographical line, which could be considered nearly as accurate as could be laid down during that time, across the continent in almost vertical direction, was available to calculate and construct other coordinates in the region while mapping the area. Most of the principal points observed lay between Agra and Poona.[71] Even after about 30 years, this survey was cited by the authorities as an example of 'minuteness and perspicuity'.[72]

Another political mission was sent in 1778, from Cuttack in Orissa to Nagpur under A.K. Elliot, for the purpose of negotiating a treaty with the Raja. The aim was to get his permission for the safe conduct of Colonel Goddard's army marching from Kalpi, on the Jumna, passing through his territory on way to Bombay. As usual with such missions, Captain William Campbell was ordered to keep up the traverse of the route of the mission. A memoir/field book[73] covering his route from Cuttack to Hoshangabad is preserved in the National Archives of India, New Delhi. The survey could not be properly carried up to Nagpur by Campbell, as he fell ill. Despite the death of the leader of the mission, A.K. Elliot on the way, the party proceeded to Hoshangabad where they were to meet Colonel Goddard who was halting there. They reached Hoshangabad on 21 December 1778. Thereafter, the survey observations appeared to have been made intermittently, as Campbell himself was in ill health. The memoir gives an excellent idea of the mode and nature of survey that he undertook on the route. Many compass-bearings, directions of the flow of the rivers and streams, coordinates of hills, together with the notes regarding the nature of topography of the region traversed and even the time of passing through certain villages are noted. Distances between two points, were, at times, calculated with the help of their average calculated travelling speed. The actual protraction of this survey was made some years later, probably, by R.H. Colebrooke.[74] As disaster had fallen upon the mission on its way, the political mission could not achieve its purpose as the leader, A.K. Elliot, died near Sarangarh before reaching Nagpur on 15 November 1778, but the survey journal,[75] with the National Archives, ends on 21 December 1778. The route followed was from Cuttack, via Boad, Sambalpur, Rattanpur, Khairagarh on to Nagpur, the capital of Berar.

During the closing years of the first Maratha war, Warren Hastings, the Governor-General, sent a political mission to the Court of the Maratha Raja of Berar under Charles Chapman, for establishing friendly relations with him in November 1781. James Ewart was the surveyor with this mission. He ran a continuous traverse survey from Chatra through Lohardaga, Rattanpur, Khairagarh to Nagpur.[76] The map, thus prepared, is also lodged with the National Archives of India, New Delhi. Later, Ewart sent maps of various routes prepared by him when he was posted at Nagpur. The Surveyor-General had reported about his activities at Nagpur in his report to the Government: '... he not only sent me his Route laid down from observation and measurement, during his stay at Burra Nagpur he was very attentive in making astronomical observations and procuring me several routes from Cossids.'[77]

Ewart got an opportunity to return via another route—he returned to Benares through Mandla, Khairagarh, Rattanpur and Surgoja to Chatra. Ewart's route surveys were very much appreciated by James Rennell.[78]

The Company used to avail itself of similar opportunities to survey the routes when Residents were appointed to the courts of Indian rulers. In 1788, George Foster was sent to Nagpur as Resident at the court of the Raja of Berar. From Kalpi, J.N. Rind was ordered to accompany him as a surveyor to survey and prepare a map of the routes.[79] In a similar fashion, James Davidson accompanied Foster from Cuttack to Nagpur in 1790 and once again, the route from Cuttack was surveyed.[80]

Apart from gathering geographical information, the Residents or other officers attached with them also collected maps, etc., of local origin. Natives were also employed for minor survey work like calculating or measuring distances from one point to another. Rind, while working as Assistant to the Resident at Delhi during 1785-7 collected sufficient information in this manner and was able to make maps of the Punjab—'Sketch map of the country of the Sikhs. Scale 1 degree to 42 *Koss*, 1789';[81] 'Routes from Agra to Delhi and from Delhi to Lahore and Attock including the Punjab to Moultan, scale 1-inch to 5 $\frac{1}{3}$ miles, Surveyed by Lt. J.N. Rind, 1789 MS',[82] etc. James Rennell also mentions W. Kirkpatrick having sent him registers showing measured distances from one place to another undertaken by the orders of many of the Mughal Emperors, Akbar, Shahjahan, etc. These were found at the Emperor's Court by Kirkpatrick when he was stationed at Delhi. The registers contained distances, though without compass bearings, or directions of places situated on the major routes, leading, for example, from Lahore to Kabul, Ghazni, Candahar, Multan, etc., and also from Lahore to Kashmir and Kashmir to Attock. Latitudes of some places were also given.[83]

We have seen how the English expanded their frontiers in the plains of the north-west of Bengal up to Delhi on the banks of the river Jumna and to the south-western borders of Nepal by the end of the eighteenth century. The entire territory of Rohilkhand and the ceded Districts of Oudh comprising the districts of Allahabad, Fatehpur, Kanpur, Azamgarh, Gorakhpur, Bareilly, Moradabad, Bijnour, Budaun and Shahjahanpur had become part of their possessions by 1801. Thus the English now touched almost the entire southern boundary of Nepal. The survey of their own territory was no problem.

Earlier, the geographical information about the regions bordering Nepal, Tibet and Burma was gathered mostly from the letters written by Jesuit Missionaries going to Tibet or China. In 1624, Antomio de Andrade, who came to Goa in 1600, travelled from Agra to Srinagar in the Garhwal region and then over the Mana Pass to Tsaparang on the Sutlej river and back to Agra.[84] In 1631, Francis de Azeveda followed the same route to Tsaparang, on to Leh and by way of Lahul and Kulu, came back to Agra.[85] Then, in the eighteenth century, Father Ippolito Desideri, travelled into Kashmir including Leh in Ladakh.

During the second half of the eighteenth century, two Frenchmen helped the cause of the Indian cartography of north India. The first was Jean Law de Lauriston. After the defeat of the French in 1757, with a small body of loyal French and Indian soldiers, he moved from place to place. In his wanderings, he visited Lucknow, Delhi, Agra, Bundelkhand, etc., and ultimately surrendered to the English, who finally deported him to France. The surveys of his marches, along with his map, he gave to d'Anville there, who got the map engraved in 1763.[86]

Another Frenchman, a scholar, Anquetil Duperron, was in India during 1757-61. He travelled from Bengal to Masulipatam, through the Orissa coast; then, later, he went from Goa to Aurangabad through Poona. He maintained details of measurements and topographical observations which were later used by James Rennell.[87]

The controversy regarding the sources of the rivers Ganges and Brahmaputra also in a way indirectly helped in the explorations of the regions like upper reaches of Kumaon Hills and upper Assam, etc. The position of the sources of these rivers, particularly that of the Ganges, was in those days 'one of the most interesting problems in Indian geography'.

The accounts of Edward Terry, the Chaplain to Sir Thomas Roe in 1655, and that of Father Desideri in 1715, who came to India in 1712, both differ in their conclusions. According to Terry both the Ganges and the Brahmaputra flowed from a mountain which he called *Caucasus*, but as per the version of Father Desideri, Ganges and the Indus both issued out from the Kailash mountain. The only reliable knowledge of these regions, including that of Tibet and upper Himalaya, was derived from Chinese sources, through the Jesuit missionaries. In 1717, Father Jean-Baptiste Regis got hold of a map of Tibet, given to him by the Chinese Emperor, for whom it was prepared by two Chinese Lamas who had visited the area in 1711 and prepared the map, on their return. The Chinese Lamas reached a lake in the western Tibet in 1711. Its local name was Lanka-Dhe (the lake Rakas is situated just west of Lake Manasarowar). They were told that Lanka-Dhe was the source or the 'fountain head' of the river Ganges. Their report and the map were submitted to the Emperor on their return to Peking in 1717.[88] The accounts of these Lama explorers who went across the Tibet to specifically discover the source of the river Ganges helped, though indirectly, to understand the geography and topography of the region. Similarly, the topography of the area north of Bihar, Bengal and the upper Assam came to be known for the first time through the English expedition against Bhutan, in 1773 under George Bogle. The details given by him about his route of travel contained enough geographical material to make an estimate of the geography of the places he traversed. His route passed through Cooch Behar, Buxar in Bengal, Paro in Bhutan, the head of the Chumbi valley, between the lakes Kala Tsho and Bam Tsho and down the Penam river through Gyantse and Shigatse.[89] He reached the residence of Teshu Lama after crossing Tsang-po river on 12 November 1774. During 1783-4, another opportunity came in the way of the English. The second mission was sent to Tibet on the occasion of the re-incarnation ceremony of the Teshu Lama. Samuel Turner, along with Samuel Davis as surveyor, started for Tibet. They took the route earlier followed by George Bogle in 1773 through Bhutan. The only deviation they made was their visit to the capital of Bhutan, Punakha. There, Davis was not allowed to proceed further as he was suspected because of his profession. Turner, therefore, alone had to perform the long journey along with the mission officials. On his return in 1800 he published his account along with a map of the route followed by him.[90]

Turning now to the north-eastern part of India—towards the end of the eighteenth country, the Raja of Assam was in great trouble. His own people had rebelled against him and he had fled his capital and was being besieged in Gawahati. He appealed to the English in 1792 to help regain his capital and throne. The English did not miss the opportunity. A force was despatched under Captain T. Welsh. The rebellion against the Raja was crushed. Thomas Wood, who was the surveyor with the force, was able to do a good job of surveying. He knew that in 1765 James Rennell had completed his survey of the Brahmaputra river up to Goalpara only. When the war ended in 1794, he carried the survey further up to the Dikho river.[91]

Till 1800 the English had sent in their army to the north-eastern frontier only on two occasions: once, when Manipur wanted help against the Burmese in 1762; and the second time, when on the northern frontier of Sylhet (now in Bangla Desh), there was trouble caused by the Raja of Jaintia in 1774. On the earlier occasion, the force was led by Verelst, the English Chief at Chittagong, which passed through Chittagong, across the Fenny river, going through Tipperah and Sylhet. Due to heavy rains, however, it could not proceed further. On the second occasion, as the frontiers of all the states in this region were unmarked or undefined, nothing much could be done except that the English expedition against the raiders from across Jaintia, which was confined mostly to the plains, was successful and the raiders were driven back. No more area could be traversed and charted.[92]

Being a trading company, the English were initially, mostly interested in the sea coasts of India or in the navigability of the great rivers of the country as their factories were generally situated on the sea shores or on the banks of the great rivers, like Hooghly in Bengal, which is navigable to a great length. It is but natural, therefore, that the earliest surveys they carried out in India were of the coasts or coastal regions. The sea charts prepared by the French navigator Apres de Mannevillette were very helpful to the English. These charts were published by him in an atlas *Neptune Orientale*, in 1745.[93]

The effect of monsoon winds on navigation on the Malabar coast was quite different from that on the Coromandel coast. The Coromandel coast was much more turbulent especially from May to November and there were no safe anchorages on the entire length of the coast land. The English had come to know of it at a very heavy cost; they had lost many ships on the Coromandel coast during bad weathers. During this period, the weather affected the Gangetic delta as well adversely. Besides bad weather, this was because of insufficient and incorrect depiction of the coastal region on earlier maps as well. The first opportunity to make a proper and detailed survey of this part of the north-western coast of the Bay of Bengal came up in 1761 when Bengal, Bihar and Orissa came under the English control. Immediately after the cession, Bartholomew Plaisted was employed on the survey of the coastal area near Chittagong and also on the branches of the river Ganges close by. In 1765, he surveyed the Balasore coast. During this period, John Ritchie was surveying the Sundarbans area. In 1769, the Bengal Council which was full of praise for Ritchie's work, reported to the Court of Directors, passing his maps and charts on to them with a request that the same be engraved and some engraved copies be sent back to them.[94]

By now the number of ships in the fleet of the East India Company had increased considerably due to increase in trade. The complexion of the Company also changed from a sea power alone to combined land and sea power.

John Ritchie was asked in 1770 to make a general survey of the coast and the islands around the Bay of Bengal. The survey was carried out during 1770 and 1771. He started from Calcutta on 30, November 1770, taking soundings, measurements and astronomical observations, etc., proceeded towards the east coast of the Bay of Bengal, down to the Great Andaman and then to Little Andaman, Car Nicobar, reaching Nancowry harbour. Then, because of the bad weather, he moved straight on to the Madras Coast, began moving up the Coromandel coast, taking usual observations, etc., reached Balasore and Point Palmyras.[95]

In 1772, the Court of Directors asked the Madras Council to look for a suitable port for the safety of the ships during war or the turbulent season of the Monsoon, and had ordered them, in this regard, to examine the suitability of Coringa Bay for the purpose.[96]

To afford protection to factories established on the Indian soil, which were mostly on the sea shore, or on the banks of the river estuaries, and also to defend their own fleet from the Portuguese aggression and the pirates who were, in those days, frequenting the Arabian Sea, a nucleus was first formed on the western coast. A few fighting vessels were collected at Bombay. This developed first into the Bombay Marine or Indian Marine, and then into the Indian Navy. The Bombay Marine was initially formed in 1613.[97]

Apart from the work of protecting the English mercantile and maritime interests during wartime, the Bombay Marine also took up the job of marine surveying. Incidentally, this marine surveying and their resultant charts came in handy to give India a proper and accurate shape. As already described, for easy and rapid movement to their factories through the waterways and also to locate and mark the water hazards, like hidden rocks, etc., that used to come in their way, preliminary explorations of the coasts of India and that of other countries which fell on their way to India, or where the English had their commercial interest, such as part of Persian shores, or part of Arabian coast falling between Muscat and Ras-u-hadd, Makaran and Sind (now in Pakistan), were undertaken[98] in 1772, under the command of Lieutenant J.J. Robinson, by Porter, Archibald Blair, and Joseph Mascall. 'The Survey was minute in some parts, and when we consider the very inferior instruments with which it was conducted, its correctness is remarkable....'[99]

During the latter part of the eighteenth century, between 1770 and 1785, John Ritchie was the head of the Marine Survey Department in Calcutta. He was actively engaged in surveying the coast around the northern part of Bay of Bengal from Ganjam to Chittagong (1768-70); continued it up to Arracan in Burma during 1770-1, and by 1772, he had covered Andaman and Nicobar islands, including Nancowry harbour.[100] These surveys proved very useful to James Rennell in the preparation of his *Map of Hindostan*. Many of his other charts were later engraved by Alexander Dalrymple, the Hydrographer to the East India Company [101] In 1773, W.A. Skynner surveyed the mouth of Broach river, Gulf of Cambay and also part of the Kathiawar coast.[102]

In compliance with the wishes of the Court of Directors, the Madras Government had asked William Stevens to survey the region as well as the Bay of Coringa. In 1773, on the completion of the survey by Stevens, his chart of the Bay and Harbour of Coringa was sent to London. Then ensued a long correspondence on the subject with the Court of Directors, which dragged on till 1783 when the Court finally ordered the Madras Council to survey the entire coast from Madras to Masulipatam. In the absence of any proper arrangement for marine survey available with them, the Madras Council requested the Bengal Government to spare the services of John Ritchie.[103] But he could not be made available as he was busy surveying the Orissa coast and the mouths of the rivers of the region.

On 28 February 1778, Ritchie sent in his report of the survey of Point Palmyras and False Point and of part of Orissa coast, describing the geographical features, tidal waves and their effect on the navigation in the Bay of Bengal, and similar other navigational hazards in the area to the Bengal Council for onward transmission to the Directors.[104] Meanwhile, the Madras Council not getting Ritchie, put Michael Topping on this duty. As there was no survey ship available to survey from the sea side, he had to carry out the survey by land based on triangulation. This was perhaps the first Indian sea survey based on triangulation[105] on land which took off from a baseline measured at Porto Novo in May 1788. The triangulation covered the coastal area from Madras to Adirampatnam which lies on the northern shore of Palk Straits. The survey of the coast began from near Pondicherry southward. The actual surveying was taken up in August 1789. The map of Bay of Coringa, along with its report, was sent to London in February 1790.[106] During this period Michael Topping had made several survey trips to the region. He surveyed the mouths of the Godavery river and Coringa roads in September 1790 and January 1791.[107] These surveys were put together in the form of a map by his assistant, J. Goldingham in 1796 entitled—'Survey of the Western Coast of the Bay of Bengal from Adirampatnam to Coringa; also of part of the Rivers Kistnah and Godavery. Scale 1' = 5 miles, 1796'.[108]

In the meanwhile, the Court of Directors wrote to the Bengal Government to improve upon the charts for the security of the navigation by seeking the cooperation of every person who was under the protection of the East India Company and conversant in the naval affairs. This was done through a general public notice issued by the Surveyor-General. It was decided to receive such information that may help in the improvement of the charts that were being commonly used by the persons interested in naval affairs.[109] This public notice was the result of the efforts of Alexander Dalrymple who was an enthusiastic scholar of marine surveys and had been examining for the last many years the nautical charts and journals lodged in the India House and as a consequence of his Memorial to the Court of Directors dated the 28th January 1779. He was made the first Hydrographer to the East India Company in April 1779.[110] He published many valuable charts and journals.

Between 1777 and 1795, Archibald Blair at that time working independent of Lieutenant J.J. Robinson, was busy making surveys of the coast of Kathiawar, Salsette and also the Andaman Islands.[111] In December 1788, Blair was sent on to survey the Andaman islands and was issued specific instructions for the purpose

> ... The material object of this survey is to ascertain in what parts of the Islands there are good harbours, and where it would be most for the Company's advantage to possess one.... It appears that the most advantageous situation for an Harbour must be near the South end; and to the Eastward of the island.... It is therefore wished that you should make the first examination on this quarter; the Board are further encouraged to give you such advice, from a perusal of reports from Mr. Ritchie....
>
> The Primary view of this research being ... the acquisition of an Harbour where fleets in time of a war can refit ... on leaving the Coast of Coromandel upon the approach of the stormy Monsoon, ... the following objects occur as necessary to be enquired into: ...
>
> As minute a description as time and circumstances permit to be made of the adjacent Heights, if any, and Ground, the General surface of the Ground....
>
> It is hardly necessary to recommend to you to ascertain from Astronomical observations, by such instruments as you possess, the position of the places you visit....[112]

Blair submitted his full report on the Andaman surveys on 10 June 1789.[113] Alexander Kyd and Robert Colebrooke also followed Blair to the Andaman Islands with the squadron of Commodore Cornwallis who were also officially surveying islands and coasts of the Bay of Bengal. They reached Port Cornwallis on December 23, 1789 in the morning. Colebrooke's famous journals give interesting account of the geography of the entire group of islands and description of the people inhabiting these islands.[114]

The full report of Archibald Blair was extremely well commended by Dr. Mouatt, the leader of 1858 mission to Andamans, who said while reporting on these islands, that—not only did it contain a minute and accurate account of the survey conducted under the superintendence of that able officer (Blair), but it is illustrated by a chart, in which the situation of the most remarkable localities was distinctly marked, accompanied with a plan of three harbours, which he had found to be sure places of refuge for the shipping that stress of weather or other causes might drive on the Andaman coasts. The report merited and obtained much praise for the clearness with which it was written, and the intelligible manner in which various operations of the surveying party were described. The chief geographical features of the island were delineated with a fidelity that has secured the approbation of subsequent explorers.[115]

We have had a look at almost all the major marine surveys conducted by Company's officials during the seventeenth and eighteenth centuries of a portion of Indian coasts and islands around the Bay of Bengal. We may now have a look at the surveys of the western coast of the peninsula.

We have already noticed the surveys of Kathiawar coast and Salsette by Archibald Blair, during 1777 and 1795.[116] In 1786, East India Company ordered the Bombay Government to survey the areas off Bombay harbour and the entrance to the Gulf of Cambay. Mc.Cluer of the Bombay Marine was asked to do the job. With only one assistant, Lieutenant John Proctor, he began this survey on 12 October 1787.[117] The survey of the northern leg from Bombay harbour was initially hampered by strong winds on that part of the coast. Ultimately, by September 1788, he completed the survey of the coast from 17° 12' N to 21° 40'N, including the Gulf of Cambay and the southern Kathiawar coast up to Diu.[118] He made a chart of the Bombay harbour as well with the help of Lieutenant Court. This chart is considered to be one of the earliest plans of the Bombay harbour; yet another assistant, Lieutenant Wedgeborough, surveyed and made charts of Laccadive islands.[119]

During the eighteenth century, the Arabian sea was very turbulent. Navigation was not quite safe. Pirates and others were at large, looting unarmed ships. Ragoji Angria was very active there and was working against the English interest on that coast.

From 1790 to 1793, Mc.Cluer began a systematic survey of the west coast with the assistance of Lieutenants John Ringrose, W.A. Skynner and Wedgeborough, all of the Marine. His original idea was to conduct the survey from Bombay down to Cape Comorin, but somehow he could not complete it up to the Cape, and part of the southern-most coast was finally mapped from the surveys of Captains Charles Montrion and Selby.[120] James Rennell, however, recorded that Mc.Cluer could survey only up to Angengo,[121] in the southern coast of modern Kerala. The charts of these surveys by Mc.Cluer were prepared by Lieutenant Wedgeborough, one of his assistants. The final completion of the charts of the entire western coast was achieved by Lieutenant A. Dundas Taylor in 1853.[122] Apart from these surveys Mc.Cluer compiled from his own observations extensive tables of latitudes and longitudes, and wrote his own personal notes on these surveys. Some doubt prevailed at that time about the accuracy of his surveys northwards from Bombay. Captain T. Jervis, the Surveyor-General, on the other hand, in his 'Report on Surveys', stated:

I should not omit to notice the valuable maritime surveys of Captains Huddart and Mc.Cluer, and Lieutenants J. Ringrose, Wedgeborough and Skinner, on the western coast of India, from 1790 to 1793, which still continue to be good authority to navigators of that coast, and were actually incorporated by Colonel Charles Reynolds in his map. (This map of Colonel Reynols was, however, never published) .... [The outcry]; which subsequent inquiry has shown to be without a shadow of justice; and I may mention it as a corroborative proof of the attention and skill which must have been bestowed on the subject by Captain Mc.Cluer, that in carrying on a trigonometrical and geographical survey of the coast upwards,... I found the actual outlines of the coast and exact distances differ very immaterially from these in Mc.Cluer's charts, and I had the more favourable opportunity for verifying the fact, as the Superintendent of Marine furnished me with Captain Mc.Cluer's original drafts, on a large scale, for the express purpose.[123]

Thus, almost the entire sea coast of India beginning from the present Pakistan on the Arabian Sea side in the west, to the western and eastern coasts of the peninsula, the entire Bay of Bengal from Gangetic delta to Chittagong and Arakan coast, was surveyed and delineated by the end of the eighteenth century. This provided us with almost the exact shape of the peninsula and the Gangetic delta.

Obviously during military operations or even reconnaissance, detailed river surveys were not possible. The method of survey at that time was quite different. We can have an idea of this from what Captain P. Don had to say in 1799, while surveying the routes with General James Stuart's army in Oudh and Rohilkhand. He stated that they have to

notice all passes or defiles—whether formed by marshes, jungles, nullahs—Their nature, whether practicable for artilly [Artillery] & cavalry—Their length & Breadth ... villages whether passable by Artilly [Artillery] and cavalry—one or more roads through them ... the Nature of the Country—open or enclosed—broken or plain—cultivated or not—Tanks and Hills —Topes—jungles—Pagodas ... Forts—strength size and situation—Rivers, their Barrages [Barrags] and Fords—the Breadth of the Former—the situation of the latter at difft [different] seasons ... the Nature of the banks and bottom ... marshes, Ditches, Enclosures—on a large plain—Ground for Camps to be noticed and when—mily [military] positions occur ... where a river or deep Broken Ground for one Flank and any favourable height etc. for the other—the Adjacent Ground to be examined....[124]

But the nature of survey of a territory, more especially, the river survey when the survey party was not in a hurry, and there was no outside authority to harass them or interfere with their work, or when the course of the river to be surveyed fell in the region controlled by the surveying power, detailed topographical survey could be undertaken. This was the case with the surveys, during the seventeenth and eighteenth centuries in Bengal, Bihar and Orissa and to some extent in Oudh and Rohilkhand and other parts of India described hereafter.

Many of the earlier surveyors had marked the courses of the rivers not from exact surveys, but from compass readings, visual observations and to some extent by measurements. But during the period under review, under more favourable circumstances, the method of the river survey could be ascertained or understood from the detailed written instructions of Robert Colebrooke, the Surveyor-General, to Hoare, who was to survey the Jumna from Allahabad to Delhi in 1795. Colebrooke used to issue detailed written instructions to every surveyor and was in the habit of scrutinising the field books submitted by the surveyor when the work was partially over, and giving his remarks for corrections, etc. To Lt. James Hoare, he wrote

I would advise you, so far as the river may be navigable, to travel by water, with a perambulator driving along the bank and keeping pace with your boats, when you arrive at any town, fort, or Gaut [ghat], it will be easy to step out of your boat and look at the wheel, and the intermediate distances may be known by a time-keeper or a good watch, which if your boatmen are made to keep an even pace, may be calculated by the rule of proportion. I would, however, advise you to observe the wheel as often as possible, and for the greater accuracy of your survey to proceed slowly, and not to be over-anxious to make long journeys.

The direction and bearings of the River may be ascertained with sufficient exactness by a compass; but a Theodolite will be necessary for nicer observations on shore. If you can also observe by Astronomical observations the Longitudes of two or three principal places, your work will be complete.[125]

To understand the detailed mode of river survey, we may examine the criticism offered by Colebrooke on the receipt of the first field book of Hoare

The specimen transmitted by you appears in some respects sufficiently satisfactory, the distances being marked with minute preciseness; but I beg leave to observe that all the Angles being only given to points of the compass, it will be difficult for you to project your work without running into considerable error.

I beg leave to remind you that in my instructions I mentioned that a compass might be sufficient to ascertain the directions and windings of the river, but that a theodolite would be necessary for nicer observations on shore. It does not, however, appear that you had used one, but even with an Instrument of the former kind you might have marked the bearings of distant objects with greater precision, and by observing from time to time the Sun's Amplitude or Azimuth, the variations of the needle might have been found, so as to render your sights sufficiently correct for the common purposes of Geography.[126]

Because of the location of the English factories the English were, in the beginning, more interested in keeping their communications open, and as such their first priority in comparatively peaceful situations was the

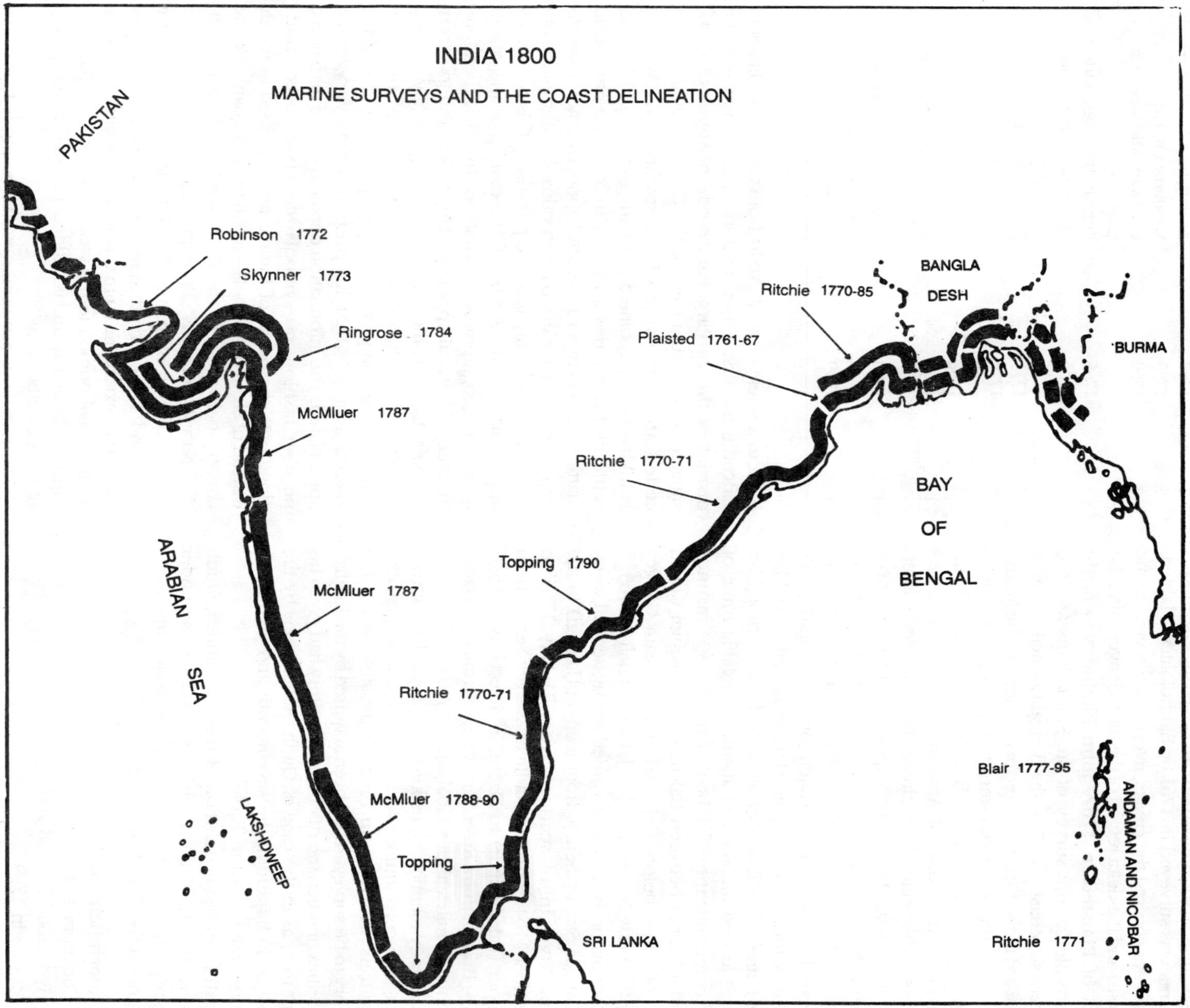

Map 19: *India, Marine Surveys and the Coast Delineation* (AD 1800). (Compiled by the author from various sources.
Note: Indian coast line has been photographically enlarged from the Survey of India 1:21 million Plain Outline Map of India, Printed 1987.

survey of the sea coasts and the river approaches. Their factories at Dacca and Chittagong were very hazardous to approach by open sea and therefore finding out a safe passage through the Sundarbans was given the top priority. This area was surveyed in 1761 by Bartholomew Plaisted. His surveys covered the creeks, rivers, islands, etc., on the sea coast and the delta region of the river Ganges. After the death of Plaisted in 1767, John Ritchie was made responsible for the survey of the Sundarbans and the adjacent sea coast. By 1771, the Bengal Council was able to forward to the Court of Directors the map of the Mouth of the Calcutta River (western branch of the Ganges) along with a Chart of the Bay of Bengal from Point Palmyras to the Coast of Arracan (Burma).[127] Ritchie's exploits in this region have already been detailed earlier.

For many years before the acquisition of the provinces of Bengal, Bihar and Orissa, the waterways up the Ganges and its tributaries including Brahmaputra were the main lines of communication for persons and goods to their factories and for internal trade as well. On 6 May 1764, James Rennell was ordered by the Governor, Van Sittart[128] to find the shortest all-time navigable route of river communication for boats carrying 300 maunds (about 11 tons) between Calcutta (via Channel Creck, some 60 miles below Calcutta) and the main Ganges River.[129] This was completed by James Rennell in 1765.[130]

Other major river surveys carried out by Rennell were, (1) Ganges Survey Series, from Jelinghi (Jalangi) to the Megna River in 14 sections, from 21 May 1764 to December 1764;[131] (2) Megna River Survey Series, from Lakhipur (Lakshmipur) to Dacca, in 5 sections, 1765;[132] and (3) Brahmaputra River Survey Series from near Dacca to about 20 miles above Goalpara, in Assam, 1765.[133]

As already mentioned, making of a proper land survey was a time-consuming process; and required peaceful conditions which could only be possible when the territory to be surveyed had been acquired by the English authorities or when the ruler of the territory had given his assent to it, as in the case of Oudh. Acquisitions by the British in India during the seventeenth and eighteenth centuries are shown in Table 1.

TABLE 1

Sl.No.	*When acquired*	*Territory acquired*
1.	1757	24-Parganas
2.	1759	Masulipatam, etc.
3.	1760	Burdwan, Midnapore & Chittagong
4.	1765	Bengal, Bihar and Orissa
5.	1765	*Jagir* land near Madras
6.	1766	Northern Circars
7.	1775	Zamindari of Benares.
8.	1776	Island of Salsette
9.	1778	Nagore
10.	1778	Guntur Circar
11.	1792	Malabar, Dindigul, Salem, Baramahal, etc.
12.	1799	Coimbatore, Canara, Waynaad, etc.
13.	1799	Tanjore

Apart from the acquisition of these territories, Oudh became an ally of the British, and Arcot came under British protection. The rulers of the following states recognized the British supremacy:

1. Kooch Behar in 1772
2. Cochin in 1791
3. Hyderabad in 1798 and
4. Mysore in 1799

We may now enumerate in some detail the land surveys of territories which were more important militarily, commercially and in terms of land revenue. These were the earliest to be properly surveyed and would enable us to begin to understand the geography of India in a scientific way. Apart from the narrow lands of Calcutta, the first major territorial acquisition was the district of the 24-Parganas (modern Alipur district), situated in the south-west of the Gangetic delta. It was ceded after the battle of Plassey in 1757. William Frankland was appointed as its surveyor to find out and settle its boundaries and also its revenue and other matters. In December 1758, after about five months of his appointment he collected much valuable information but it was useful for only revenue purposes. This was much appreciated by the Court of Directors, but the report lacked in geographical information. In 1761, therefore, the Bengal Council appointed Hugh Cameron to survey the New Lands as the acquisition was generally termed. Cameron's designation was 'surveyor of the New Lands'.[134] He completed the survey in 1761-2, and the skeleton map[135] that he prepared was very much appreciated. The information depicted on the map is in a peculiar fashion; no symbols have been used; lands along both the banks of Jumna are described in words instead of signs or symbols—as on the right bank, 'A fine country belonging to the Company'—and on the left, 'The Nawab's Country', both the banks, thus marked, are devoid of any information about topography. During 1767 and 1770, Claud Martin surveyed the southern *parganas*, extending from Calcutta to Sundarbans, along with a small portion to the west of the Hughly. This is a topographical map and is entitled 'Part of General Survey of the Calcutta Lands'.[136]

The next land acquisition worth noting is that of the districts of Midnapur, Burdwan and Chittagong (now in Bangladesh). Chittagong was surveyed by Bartholomew Plaisted and William Richards.[137]

By 1760, when the districts of Midnapur, Burdwan and Chittagong came into the Company's possession, very little geography of Bengal was known. This is evident from the examination of the map of India prepared by J.B.B. d'Anville in 1752. d'Anville was known to have based his map on the information that he received through 'authentic sources' only, otherwise he preferred to leave the area blank. The map, in question, with the exception of some place names on the banks of major river courses does not show any topography and the Bengal region looks almost blank. The opportunity to start a proper survey came their way only when the English took possession of the provinces of Midnapur, Burdwan and Chittagong in 1760. No specific maps of Midnapur and Bardwan of the time were available, except 'Knox's Roads in the Midnapur Province' made by James Nicol in 1761-3.[138] Whatever other geographical information that was available was through route surveys of the army, marching through the region, already detailed earlier.

The Northern Circars was the region comprising the circars of Kondapalli, Ellore, Rajahmundry, and Chicacole; or from the northern banks of Kistna river, about 17° N latitude to the southern shores of Chilka Lake, about 20° N latitude, running approximately along the northern half of the Coromandel coast. Although their forces had entered the Northern Circars earlier also while fighting the southern peninsular wars, in 1765 the English were granted the *Diwani* of all these Circars by the Mughal Emperor at Delhi and finally the Northern Circars were ceded to the English by the treaty of 1766 with the Nizam, except the Guntur Circar which was acquired in 1778. But the occupation of the Northern Circars was hindered by the war with Haider Ali in 1767. General J. Cridland advanced into the Circars north of the Kistna river while Colonel Peach moved up through Ellore and reached Warangal and threatened Hyderabad. A new treaty was signed in March 1768, whereby the Nizam confirmed the cession of the Circars. Colonel Peach's force went up to Ganjam.

As was usual with the English army, route surveys were conducted by the surveyors of the army. R. Gardner, under Colonel Peach surveyed the route from Masulipatam to Warangal; J. Cridland surveyed the Tekkali District[139] of Chicacole Circar. In 1766, Cotsford prepared a plan of Chicacole Circar and a sketch of the town of Ganjam;[140] in 1769, he made a Plan of the Itchapour (Ichehapuram) District.[141] Between 1767 and 1775, W. Stevens, P. Pittman, Dugood, Cotsford and James Johnston contributed towards the preparation of boundary survey maps of the boundary line with the Nizam.[142] In 1783, a detailed survey of the country between Ongole and Cuddapah was made by Walter Lennon.[143] This involved taking astronomical observations of some places, the distances measured by perambulator, bearings taken from Azimuth compass, etc. His surveys were used for revenue purposes as well. In 1788, Lennon surveyed the territory between Rajahmundry and the Godavari river.

Next came the survey of the provinces of Bengal, Bihar and Orissa, acquired by the English in 1765. Before discussing the official maps after the appointment of James Rennell as the 'Surveyor of the New Lands',[144] succeeding Hugh Cameron on 9 April 1764, important maps of Bengal by some individual cartographers, compilers or publishers may be noted.

The first such map that came to light was the one compiled by Jean de Barros, *'Describcao do Reino de Bengalla'*, *c.*1550. This Portuguese map is east-oriented and covers from *Orixa* (Orissa) in the south-west to Arra (Arakan) in the south-east and *Patanes* (Patna?) in the north-west to *Lima* (Lushai region?) in the north-east.[145]

Another map of Bengal, worth mentioning is discussed in *Bengal Past & Present.* It is also a Portuguese map of *Bengal and Arracan*. During seventeenth and eighteenth centuries, this was considered one of the accurate maps of the region. The map was prepared by Mattheus van den Broeck, the Dutch Director of Trade, Bengal, during 1658-63 and a member of the Council, Dutch East Indies, 1763-9.[146]

Two old maps of Bengal are also listed in the old Catalogue of Maps & Plans in the Imperial Library at Calcutta, 1910; (1) *Sinus Gangeticus vulgo Golf de Bengála*, and (2) *Royaume de Bengala et les Pays voisins du Gange, situentre le Mogolistan et le Pegu.*[147]

At the time of the acquisition of the provinces of Bengal, Bihar and Orissa in 1765, the Bengal Government was so devoid of good maps of Bengal that the Court of Directors, on the demand of the Bengal Council sent to them from London some detailed maps of southern India and Bengal—'We send in the packets by those ships, agreeable to your request, some of the best maps we can procure of the Peninsula of India; likewise some others lately printed of the Bengal Provinces and Countries adjacent.'[148]

The *'Map of the Kingdom of Bengal'*, based on the surveys of James Rennell, the details of the surveys of the map being carried to England by Henry Vansittart, late Governor of Bengal, 1760-4 was compiled by Samuel Dunn in England. It covered the regions from the Ganges down to the Bay of Bengal near Balasore in Orissa, and from Hughly in the east to the Rajmahal hills in the west. This map was probably included in the packets.[149] Presumably, a reduced form of the huge map of part of Bengal, 5 feet by 3 feet in size, and drawn on the scale of 10 miles to an inch, prepared in India and carried to England by Clive early in 1767,[150] was also one of the maps included in the packets.

Robert Orme, the historian exploited his closeness to both the Governors, Vansittart and Clive both of whom helped him with original maps of Bengal and other geographical material to enable him to compile his own

Map 20: *Map of Bengal* by De Barros (*c.* AD 1550). (Gratefully taken from J.J.A. Campos, *History of the Portuguese in Bengal,* pubd. by Butterworth & Co., Calcutta, 1919).

maps. He had settled in London and at that time it had become difficult for him to consult records, then being accumulated at the India House. He, therefore, wrote to Clive on 12 November 1764:

> You my lord, have treated me differently; and pray continue to do so. Make me a vast map of Bengal, in which not only the outlines of the provinces, but also the different sub-divisions of Burdwan, Beerboom etc., may be justly marked....[151]

In his reply from Calcutta, dated 29 September 1765, during the second spell of Governorship Clive agreed to Orme's request and wrote back:

> I am preparing plans in abundance for you. You shall have very exact charts of Bengal, Bihar, and Orissa, and of the Mogul Empire as far as Delhi at least. A map of the Ganges likewise and all the other rivers of consequence.[152]

On 10 October 1795, Clive's orders were delivered to James Rennell while he was still at Dacca to prepare a general map of Bengal as early as possible. He was allowed to have an Assistant Surveyor, William Richards. He continued with his Brahmaputra river surveys also and began, with the help of Richards, pushing the general surveys towards Bihar.[153] Later, other surveyors, De Gloss, Adams, and Carter were also attached with him.[154] By August 1766, Rennell had completed the map of part of Bengal for Clive,[155] who took it with him when he left India in January 1767. During 1767, Rennell prepared another map of the Bengal province which was sent to the Court of Directors.[156]

In 1772, W. Bolts published a map of Bengal based chiefly on Rennell's surveys engraved by Thomas Jefferys, and assisted in surveys by Luis De Gloss, Showers and Motte during 1766.[157]

It may be noted that almost all the maps that were officially surveyed, were sent to the India House, London except those that were carried personally by Governors Clive and Vansittart,[158] and others, and also those sent directly to persons in England who published them privately. F.C. Hirst, while working in London was able to examine the collection at the India House during 1910-17. Though he could not manage to catalogue them all, he was, able to publish *The Surveys of Bengal by Major James Rennell, ... 1764-1777* in Calcutta in 1917.[159] On carefully analysing the maps of Bengal, Bihar and Orissa, compiled and published by James Rennell during the eighteenth century as detailed in this and the other related publications,[160] it is evident that the following official surveyors also greatly contributed in the preparation of the maps of the region, and thus made us understand the geography of these provinces in a proper way. The contribution of James Rennell was of course the greatest as shown in Table 2.

Before leaving India, Clive created the post of 'Surveyor-General' in Bengal, and appointed James Rennell as the first Surveyor-General.[161] Clive's great regard for Rennell is clear from what he wrote to the Court of Directors:

TABLE 2

Sl.No.	*Surveyor*	*Period for which their surveys were used in Rennell's maps*
1.	Bartholomew Plaisted	1760-1
2.	Russell	1761-71
3.	Polier	1764
4.	John Richie	1764-73
5.	Claud Martin	1764-73
6.	Luis Du Gloss	1764-75
7.	James Rennell	1764-75
8.	William Richards	1764-75
9.	Camac	1766-70, 1775
10.	Carter	1766-75
11.	Ironside	1766-70, 1775
12.	Morrison	1766-70, 1775
13.	Alexander Kyd	1767
14.	Portsmouth	1767-74
15.	John Adams	1767-73
16.	Huggens	1767-73
17.	Thomas Call	1767-74

> So much depends upon accurate surveys, both in military operations and in coming at a true knowledge of the value of your possessions, that we have employed every body on this service who could be spared and were capable of it. But as the work must ever be imperfect while it is in separate and unconnected plans, we have appointed Captain Rennell, a young man of distinguished merit in this branch, Surveyor-General, and directed him to form one general chart from those already made, and such as are now in hand as they can be collected in. This, though attended with great labour, does not prevent him [from] prosecuting his own surveys....[162]

The contribution of James Rennell, who was known as the 'Father of the Indian Surveys', in the initial stages of the development of Indian cartography, was stupendous. Even when he retired and was living in England, he continued to work on the huge collection of maps at the India House, London. A chronological list of some of his well-known maps, surveyed, compiled, printed and some of them even published by him, is as follows:

1. The Ganges from the Jelenghee to the Megna. In 14 parts. 1764. Scale, 500 yards to 1 inch, MSS.
2. A particular Plan of Custee Creek, from the head of the bar, by James Rennell. 9 June 1764. MS.
3. A general sketch of the Eastern Branch of the Comer Creek, by James Rennell, Surveyor. In 2 parts, from the head near Motiapour to the Ganges. July 1764. Scale, 1 inch to 1 mile. MSS.
4. Survey of Mandapour Creek. Nos. 2 and 3, from Maleeat to the head of the Burrashe, by Rennell, Surveyor. July 1764. Scale, 500 yards to 1 inch. MSS.

5. A general map of the Megna and other Rivers, which form the nearest passage from Luckypour to Dacca in the dry season, including likewise the Creek of Chiddypour or Luricule, which affords a short passage from the Megna to the Ganges. Reduced from the original surveys by James Rennell, Surveyor. 1765. With a Plan of the British Factory at Dacca on a larger scale. Scale, 2 miles to 1 inch. MS.
6. An exact Survey of the River Megna from its conflux with the Ganges to Cassempour, reduced from the original survey to a scale of 2 inches to 1 mile, by James Rennell, Surveyor. In 4 parts. 1765. MSS.
7. A Map of Chiddypour or Luricule Creek, which runs across the country from the Ganges to the Megna, and lies in the direct road from Sunderbound to Dacca, 1765. By James Rennell. Scale, 2 inches to 1 British mile. MS.
8. A general map of the Luckya River from its separation from the Baramputrey to its conflux with the Issamuty and Megna, with Noranda Creek and the environs of the city of Dacca. 1765. By James Rennell, Surveyor. Scale 3 miles to an inch. MS.
9. A general Map of the River Baramputrey, from its conflux with Issamutey, near Dacca, towards Assam, carefully reduced from the original surveys to a scale proper for common use. By James Rennell. 1765 In 5 Parts. Scale, 2 miles to 1 inch. MSS.
10. An exact Survey of the River Baramputrey from its conflux with the Issamuty, near Dacca, towards Assam. By James Rennell, Surveyor. No.1 containing from the Issamuty to Nauldee Point ... 1765. Scale, 2 inches to 1 mile. MS.
11. An exact Survey of the River Megna and Baramputrey, by James Rennell, Surveyor. No.1 containing from the Issamuty to Nauldee Point,... May 1765. Scale, 2 inches to 1 mile.
12. A Map of the Denospour or Raagegunge (River from the junction with the Ganges) to Raagegunge, containing near 200 miles of the course of the River. Surveyed in 1766 and 1767 by Rennell and Richards. Scale, 3 miles to 1 inch. MS.
13. A Map of the great River Ganges, from Jelenghee to the junction with the Megna, including its several branches on both sides as far as they have been surveyed; together with River Megna from Luckipour to the mouth of the Issamuty, and the environs of the city. 1767. Scale, 3 miles to an inch. MS.
14. Ganges, No. 2 (river above and below Colgong). Surveyed in June 1767. (By James Rennell, Surveyor). Scale, 2 inches to 1 mile. MS.
15. (First Series). A set of general and particular Maps of Bengal and Bihar, with general maps of Allahabad and Awd. Drawn from actual surveys taken between the years 1763 and 1774, by ... East India Company's Surveyors. Constructed from the original surveys by James Rennell, Surveyor-General. MSS.
16. A Chart of the Head of the Bay of Bengal from surveys taken in the years 1768, 1769, and 1770, by John Ritchie, Surveyor. Drawn by James Rennell, Surveyor-General. With a plan of Bominy Harbour (Fenny River) on the coast of Chittagong, 1770, enlarged. Scale, 5 miles to 1 inch ... MS.
17. (Second Series). A set of general and particular Maps of the Bay of Bengal. Drawn from actual surveys taken in the years 1768, '69, '70, and '71, by John Ritchie, Surveyor, and examined and compiled by James Rennell, Surveyor-General, 1772. MSS.
18. (Third Series). A set of general and particular Maps of Bengal, Behar, etc. By James Rennell, Surveyor-General, 1773.
19. A Map of Bengal and Behar, with part of Orixa. Drawn from actual surveys by James Rennell, Surveyor-General. 1775. Scale, 10 geographical miles to 1 inch. MS.
20. [Bengal Atlas]. A Map of Bengal and Behar in VIII Parts. Published ... by James Rennell, 10 November 1779. (This is the first edition).
21. A Bengal Atlas, containing Maps of the theatre of war and commerce on that side of Hindoostan; compiled from the original surveys, and published ... by James Rennell, ... Surveyor-General in Bengal, 1781. (This is the second edition of James Rennell's Bengal Atlas).
22. Part of Calymere Reef, by James Rennell, 1763-4. [This map was published by Alexander Dalrymple in 1782, and is a part of his collection—'A Collection of Charts, Plans and Views', Vol. I].
23. Bay east side of Camorta Isle, [Nicobar Islands], by James, 1762. [This map forms part of Alex. Dakrymole's 'A Collection of Charts, Plans, and Views'. 1782. Vol. II] We may now describe the official surveys taken up during the seventeenth and eighteenth centuries.

Districts of Baramahal, Dindigul, Malabar and Salem were acquired after the Treaty of Seringapatam of 1792, and Coimbatore, Kanara and Wainad districts after the fourth Mysore war of 1799. These, along with other acquisitions, during the period were very significant because the boundaries of the English territories now extended from the Coromandel to the Malabar coast.

Immediately after their accession in 1792, the districts of Baramahal and Salem were put under the charge of

Alexander Read. He made his first survey in a hurry so as to form an idea from the map[163] as to the geographical position of the roads, rivers and places covering the area. He created a record in the history of Indian surveys by using plane table survey system in the process which was based on some sort of graphic triangulation. Later, John Mather, the surveyor, was asked by Alexander Read to make a thorough survey of the districts. The work began in January 1793, and was completed by 1797.[164] During 1793 and 1794, James Allan prepared 'Military Survey of the Baramahal and Ceded Countries'. This was a rough reconnaissance of the roads.

The survey of Dindigul district was put under the charge of newly trained surveyors from Madras Surveying School, Charles Webb, Robert Gardner and J.A. Johnson in December 1796. Later, Charles Webb and R. Gardner, who died in 1798, were replaced by Thomas Turnbull and James Allan. They completed the survey by the end of 1801.[165]

Malabar given to the Bombay government by the Central government to administer was surveyed mostly during the third Mysore war in 1790. A Bombay army detachment under Lt. Colonel James Hartley was sent to the Malabar coast. Charles Reynolds surveyed the marches of the detachment through the country. During this war, he surveyed Cochin, the Malabar Coast, Calicut, and made geographical survey of Malabar as well, showing the limits of the Coorg country.[166] In late 1792, G.L. Emmitt and J.A. Johnson were ordered to survey the Malabar coast, so as to complete the survey of the entire frontier of the Malabar state under the supervision of Major J.C. Sartorius. By March 1793, J.A. Johnson was busy in south Malabar. B. Moncrieff and Woodington were helping G.L. Emmitt in north Malabar, which included Coorg, Wynad and Mount Della (Dilli) with the coast as far as Nelisuram.[167]

In December 1794, the completed maps under Emmitt were sent to Bombay.[168] In October 1795, Moncrieff replaced G.L. Emmitt and was ordered to survey south Malabar more thoroughly. After the final defeat and death of Tipu, he completed the remaining part of the survey in the north Malabar also. The acquisitions of 1799 were handed over to the Corps of Guide under Thomas Sydenham for district-wise survey.[169] With this, all the major territorial acquisitions were surveyed either topographically or by route surveys. Revenue surveys conducted in the early years were actually topographical surveys in every sense of the word.

By the end of the eighteenth century the East India Company had acquired large areas to administer—politically, socially and to collect revenues. There were military route surveyors, geographical and topographical surveyors and, in a way, revenue surveyors also. There were no separate cadres as such, and the duties in all these spheres, often overlapped. Generally, all the surveyors came from the military.

It was in 1790 when the Company had taken over the revenue administration of the districts of the Carnatic that Michael Topping had proposed the establishment of a new cadre, Assistant Revenue Surveyors, to be professionally trained at a training school, which was being established also at the suggestion of Topping. The Assistant Revenue Surveyors were to work under the District Officers. The cadre was to be headed by an Inspector of Revenue Surveys.

After the death of Topping, he was succeeded by J. Goldingham. He was also to act as Inspector of Revenue Surveys. The revenue surveyors were initially to prepare a topographical map of the area or the district in question and then to add to these maps such information as was found useful for revenue purposes like cultivation and irrigation. Thus, the maps prepared for revenue purposes could as well be studied for topographical requirements.[170]

During the last years of the eighteenth century[171] the Assistant Revenue Surveyors, surveyed or assisted in the survey of the districts of Dindigul (1796-8), the *jagir* land of Chingleput district (1797), Madura (1797) and Vishakhapatnam (1799)[172]

The end of the eighteenth century marked the end of one epoch of map-making in India. Before we go in to the developments that took place in the nineteenth century and later, we would do well to look at the major contributions of the prominent surveyors of this century. This will help us to properly assess the achievements of the surveyors.

In 1767 the post of Surveyor-General was created in the East India Company under the following Resolution of the Council, appointing James Rennell as the first Surveyor-General:

> Mr. James Rennell having, in the surveys which have lately been carried on under his directions, given sufficient proofs of his abilities and assiduity in that branch, which may prove of great consequence to the Company's possessions under this Presidency. It is agreed that he be appointed Surveyor-General, with the rank of Captain, ...[173]

The contribution of Rennell and his Assistants in the cause of Indian geography and its evolution has already been detailed earlier.

By 1777, when Rennell left India, he was able to satisfy both the Bengal Government and the Court of Directors, regarding their needs of geographical maps of Bengal, Bihar and Orissa.

About the maps of the southern peninsula of India, the situation was not as good as in the north. Military reconnaissance and route surveys were being continued as before and fresh geographical information was constantly being added to the existing knowledge of the region by the English armies.

Under these circumstances, Thomas Call was appointed Surveyor-General to succeed Rennell with effect from 7 October 1777.[174]

The following major surveys, in the comparatively peaceful atmosphere of north India, were carried out under the control of Thomas Call, the new Surveyor-General:

In December 1777, Ralph Broome was surveying the Jungleterry district[175] in 1778-83, Andrew Pringle was employed in the surveys of Subharnarekha (Swarnarekha) River and parts of Rohtas and Shahabad districts;[176] John Moulten was engaged in Rohilkhand and Oudh in 1779. His Memoirs give all the geographical and topographical details, boundaries of various parganas, irrigation facilities like wells, rivers, etc.;[177] William Hyde surveyed the Jumna river during 1779-87, and also covered the region adjacent to it. On Jumna, he concentrated from Musanagar, Kalpi, Etawah up to the Allahabad region;[178] and during 1780-1 Charles Ranken surveyed Ramgarh and Hazaribagh districts.[179]

Two main events took place during the period when Thomas Call was the Surveyor-General. These events enhanced our geographical knowledge of the central part of India along with that of north-western region of the peninsula to a considerable extent. These areas were mostly unsurveyed at that time. The first event was Goddard's march from Kalpi on the river Jumna, to Bombay to assist the Bombay Government in the war against the Marathas in 1778. The route of the detachment of the Bengal Army was from Kalpi through Bundelkhand, Hoshangabad, Burhanpur, on to Surat on the west coast. A second event of similar magnitude took place, which resulted in covering the other half of central India. A political mission under Elliot was sent from Cuttack on the east coast of Orissa to Nagpur and then to Hoshangabad, ostensibly to negotiate a treaty with the Raja of Berar. The details of both these marches have already been described earlier.

The contributions of these events to our knowledge of the geography of the region can be assessed when we get an idea of the areas covered by the marches.

The other two surveyors in the peninsula during the centuries under review, who were most active and whose contribution to the geography of the region was considerable, were Robert Kelly and John Pringle. The details of their exploits have already been reviewed earlier. Robert Kelly's collection of routes in 1783 in the Carnatic 'proved valuable guides to Sir Eyre Coote'[180] in his engagements in southern India. Another surveyor whose contribution to the Indian geography, particularly of the Carnatic, was widely appreciated was John Pringle. He compiled four books of Roads in the Carnatic; two were submitted in 1777, the third in 1785. With the third, he compiled and submitted a chart covering the routes of all the three segments of the Grand Army, engaged in the Carnatic, first under the command of Major-General Sir Hector Munro, second under Lieutenant-General Sir Eyre Coote, and the third operated under the command of Major-General James Stuart; all the three took place between 25 August 1780 and 7 June 1783. The map was later copied by J. Reichel in 1787.[181]

When Thomas Call was made Surveyor-General, he was already toying with the idea of preparing an Atlas of India and had actually started acting on it. But a peculiar situation cropped up.

But Thomas Call had already employed several Indian *munshis* and *harcaras*. Their job was to collect the required information for filling in the many gaps in his map of India. In October 1783 he wrote:

> I have for a year and half past, employed 6 munshis[182] and 30 Harcarrows[183] at my own expense, to travel through the different parts of India to collect information.... This I did with the permission of the ... Governor-General.[184]

Again in the following year, he wrote to the Government for the reimbursement of the amount spent by him on this count and enquired whether he could continue to employ temporary *munshis* and *harkaras* at the same rate as before.[185] Call had permitted his Assistants also to employ such temporary hands, if necessary. One of his Assistants, James Ewart, had employed *Cossids uasids* to traverse some routes during his stay at Nagpur in 1782-3, such as, route from Nagpur to Poona; and route from Nagpur to Hyderabad, 1783; other routes were from Nagpur to Ujjain and to Asirgarh, etc.[186]

James Rennell was active in England. He had a great influence with the Directors. It was, perhaps, on his insistence that in December 1783 Court of Directors had directed the Bengal Government: 'that copies of all charts and plans, likely to throw additional light on Indian Geography or navigation, be despatched to them...'.[187]

In response to these orders Call wrote to the Government:

> I have in hand an Atlas of India formed from a variety of materials, such as original Maps, actual Surveys, Routes, Marches of Detachments, Journals, Reports, information furnished by travellers, Histories and Voyages.
>
> This Atlas, though not finished, is in a state to convey much new and useful Geographical knowledge of this country; it will be divided into 20 sheets, afterwards the whole reduced into one portable sheet sufficiently large to exhibit all rivers, capital towns, Roads, and the grand and sub-divisions of the Empire.... It is constructed from a variety of Authorities, more or less to be depended on, according to circumstances; it would require a volume in folio to explain them and shew why I preferred this and rejected that....
>
> I am sensible to the Merit and abilities of Major Rennell, who has lately published a Map of India,... yet it will be no disparagement to him or his work, to say that, being on the

spot, I have, since his departure, had an opportunity of rendering my Map of India much more complete than his, and further that, was he furnished with all the materials I have procured, it would take him nearly as much to compose the Work as it has taken me, which would be so much time lost.

If I send home the Maps ... in the state they are, having no copies of them, I shall be unable to proceed with the continuation of them.[188]

This long explanation had an effect on the Court of Directors, and he was permitted to continue his work as he was pursuing earlier, but was ordered that no more temporary employment of *munshis*, etc., be made.

As an economy drive after the conclusion of the Mysore wars, the Bengal Government had under instructions from the Court of Directors issued orders that all the ongoing surveys be stopped and no one should be employed for surveying in future without the special orders of the Government.[189] He was also asked as to how long it would take to complete its compilation and reductions, and what would be the cost involved. He replied saying that he would require 12 months and the amount would come to Rs. 12,000.[190]

In 1786, Thomas Call got an opportunity to go back to his parent profession on promotion as Chief Engineer and resigned his job as the Surveyor-General in February 1786. To the Commander-in-Chief, he wrote:

Having completed my Map of India as far as the Materials in my possession will enable me to do, I humbly beg your permission to resign the appointment of Surveyor-General....

The Map is at present in One sheet and in a very rough state, owing to the frequent corrections it has undergone: to preserve a work that has been of so much expense to the ... Company it will be necessary to divide it into ten or twelve sheets and make fair copies of it; this can easily be done under the inspection of my successor in Office.

My motive for wishing to give up my appointment is that I may be able to apply myself to the studies of my profession as an Engineer and that I may be at liberty to pursue such other Services as my Superiors may think proper to order me on.

Captain Mark Wood is well qualified to fill the appointment of S. General he being the next Officer to me in the Corps, I take the liberty to mention him as a proper person to be appointed in my stead.'[191]

Francis Wilford, who was assisting Thomas Call, the Surveyor-General and had been working on his Atlas since 1783, continued with the same work now under the new Surveyor-General, Mark Wood.[192]

A couple of days after resuming charge as the Chief Engineer, Call wrote to the Governor-General-in-Council reporting that:

The Map of India being completed as far as the materials we are possessed of will enable me to do, for the present, I take the liberty to recommend to ... Board that a fair copy be made of it in twelve or fourteen Sheets on the same scale it is now laid down at, viz., about four inches to a degree.

The fair copy should be kept in the Council Chamber to assist the Governor-General in Council in ascertaining the position and distance of places as also to preserve a Work that has been attended with much labor and expense.

A reduced copy of the Map should be made on one Sheet on a Scale large enough to shew all Capital Towns, Rivers and Boundaries of provinces, leaving out all the minuter parts....

... the Court of Directors in a Paragraph of their Public Letter have been pleased to order Copies of the Map of India, together with all journals, Surveys, etc., to be sent to Europe—as my Allowance for Officers, Draughtsmen etc. was abolished, I found it impossible to comply with the Order and at the same time proceed with inserting the new materials I had in my possession....

The West and North West parts of India will soon be much improved by the Surveys of Captain Reynolds, Surveyor on the Bombay Establishment who has promised to send me all the Surveys he has taken....[193]

On getting orders for completing Call's Atlas, Surveyor-General Mark Wood requested the Government to write to Bombay and Madras Governments to send all the surveys that they would think worth adding to the map before it could be fair copied. After labouring hard on the map for quite some time, he reported in April 1788 about the difficulties he encountered in making the map into twelve sheets and expressed his inability to keep to the earlier promised time limit of doing so. The Surveyor-General wrote:

... It was first imagined that this work would have been completed in 12 months, and, had merely a copy of the former Plan been required, it would have been finished within that time. Independent of the unwieldy size of the map, which rendered it impracticable to be placed on any glass, on cutting it into 12 sheets, it was discovered that the paper had shrunk in many places near a twelfth part of a degree.

In the forming of the New Plan, it was by this means rendered necessary, not only to make all the projections again ... but likewise to compare the several situations and distances....

After describing the reasons for the delay of the completion of the New Plan or Call's Atlas, he referred to the addition of a thirteenth sheet, i.e. Index sheet, that would show the main 'divisions and the extent of country contained in every sheet'. He further stated:

... [the sheet] shall also contain a concise account of the Principal authorities from which the map has been compiled....

The Atlas in 13 sheets for the Council Chamber has at last been completed, and I only wait for the insertion of the authorities, and a book for arranging the several sheets to present them....[194]

Since Thomas Call had decided to resign from the post of Chief Engineer also and go home, on 24 September 1788, because of ill-health, he wrote a letter to the Government soliciting permission to take home the original reduced sheet of his map finished by Francis Wilford, instead of a copy of it.[195] In October, Thomas Call formally asked for permission to resign.[196] In his letter dated 5 November 1788, Mark Wood, the Surveyor-General, informed the Government that he has given to Call twelve sheets of 'the Grand Atlas of India' to deliver personally to the Directors in England, together with the original reduced copy of the same.[197] On 11 November 1788, he sailed from Calcutta. According to C.R. Markham Call

died soon after he landed in England; and his map which must have entailed an enormous amount of labour and expense, appears to have been lost.[198]

Recent researches indicate that he died on the voyage, but the papers and maps which were a part of his luggage, reached England safely, as reported by his cousin, John Call to the Court of Directors.[199] The manuscript copy of this beautifully drawn map, in several sheets, along with the 'Index' sheet is now preserved in the National Archives of India at New Delhi.[200]

Mark Wood's period as Surveyor-General was from 7 February 1786 to November 1788. Like Call, he also was appointed as Chief Engineer on 15 November 1788. On him fell the main responsibility of supervision of the compilation of fair copy of Call's Atlas and the preparation of the reduced copies. Like James Rennell and others he was convinced that the geographical knowledge of a region could be acquired only by systematically studying the military route surveys. On his suggestions, the following General Order was published by the Government in September 1788:

It is to be a standing regulation that all Officers Commanding Detachments of the Army, or single Corps, on a march, do keep an account of their daily movements, remarking their computed distances, the towns, villages, and rivers in their routes, the nature of the roads and places of encampments, or any other observations which they may deem material, copies of which are to be transmitted to the Quartermaster General, after the troops have arrived at their destination.[201]

The General Order also contained detailed regulations and instructions regarding the mode and type of documentation to be maintained while on survey, and later while compiling and computing the data thus collected. These requirements were further tightened with the issue of fresh orders requiring all surveyors, whether on Army route surveys or otherwise,

to transmit with their plans ... fair and correct copies of their journals or Field Books, containing all the original measurements by the wheel or chain, and every particular respecting their surveys in writing, and that no surveyor be considered ... as having fulfilled the object of his mission ... without transmitting ... such journal or Field Book.[202]

The orders and the regulations issued during the time of Mark Wood as Surveyor-General, proved very valuable in the compilation of future military route surveys and their centralisation at the Surveyor-General's Office at Calcutta, and at the India House, London, enabling the Court of Directors, the Hydrographer to the Company and the Surveyor-General to draw on them at the time of need.

The next Surveyor-General, Alexander Kyd, was not able to contribute much to the Indian cartography as he was mostly employed on service away from India. His only contributions were his surveys of Andaman and Nicobar Islands made during 1789-90, and his surveys of Mysore at the end of 1790. He submitted a list of all the maps of Andaman and Nicobar Islands to the Government in September 1792, for onward transmission to the Court of Directors.[203]

Another Surveyor, who like Thomas Call, attempted to compile a map of India, was Charles Reynolds. While attached with C.W. Malet, the English resident at Poona, he surveyed a considerable part of the Deccan. He had an opportunity to conduct survey in peninsula since 1782. It was sometime in 1787 that he was inspired to form a general map of India. But somehow he could not get the Governor-General's approval, till 1793, when he was able to convince him perhaps, by personally explaining his scheme to him and the Members of the Council and finally got his approval.[204] The Directors' sanction finally came in July 1795. They asked him to prepare a complete general map, with separate maps of each district on a large and expressive scale, with a topographical description of the country.[205] The part of the general map which Reynolds submitted in 1796 for the inspection of the Government contained north-western part of India, covering the Gulfs of Cutch and Cambay and part of the Malwa region.[206]

In 1798, Charles Reynolds was able to make a rough map in several sheets, which he laid before the Governor. He had employed many native surveyors or *harkaras* and had sent them to different parts of India to collect information. By this time not all the *harkaras* could submit their surveys. Some were still away in distant parts and could not be reached by Reynolds. He himself, single-handed, compiled the geographical information which came to him through his several native surveyors since 1795. In 1801, he was provided with an Assistant, B. Moncrieff. Reynolds could take a few sheets of his map to London to present them personally to the Directors. Twenty sheets were completed later and were despatched by the Bombay government to the Governor-General on 4 February 1809.[207] A copy of the whole map was redrawn with additions and corrections and was sent for the Government of India by the Bombay Government. One Sheet of the map—'Map of the countries North from Agra and Ajmer as high as Delhi' is preserved in the National Archives of India, New Delhi.[208]

The history of early surveys enumerated in this chapter, conducted in the seventeenth and eighteenth centuries, was not merely rough surveys, mostly carried out by the surveyors on their marches with the armies. The story unfolds the courage and determination of the

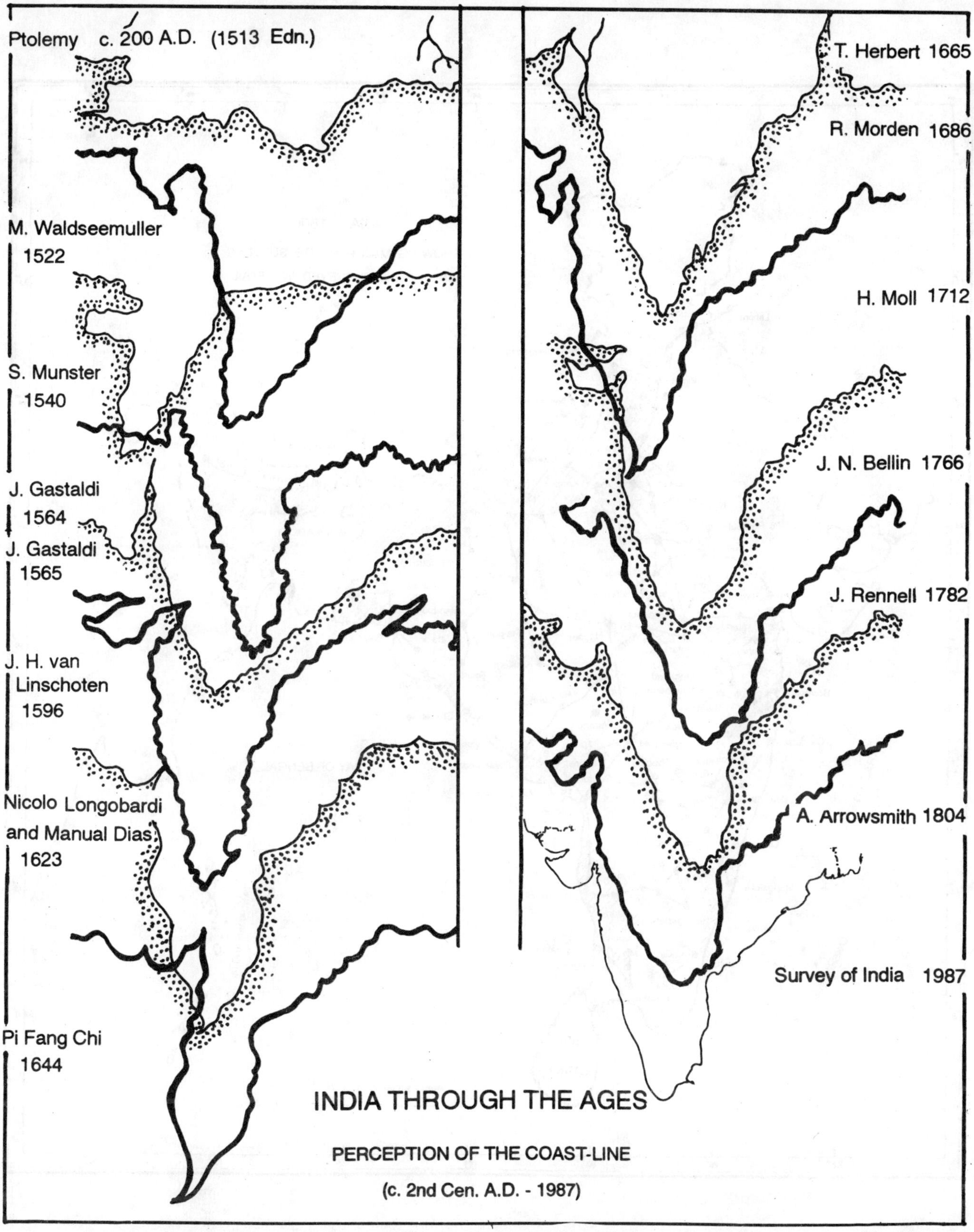

Map 21: *India Through the Ages, Perception of the Coast-line* (c. AD 200), 1987. (Compiled by the author from various sources with permission from Ms. Helen Wallis, *National Geographical Journal*, London, 1975, and Susan Gole, *A Series of Early Printed Maps in Facsimile*, New Delhi, 1984).

Note: Indian coast-line of the Survey of India 1987 map has been photographically reduced from the Survey of India 1:21 million Plain Outline Map of India, printed 1987.

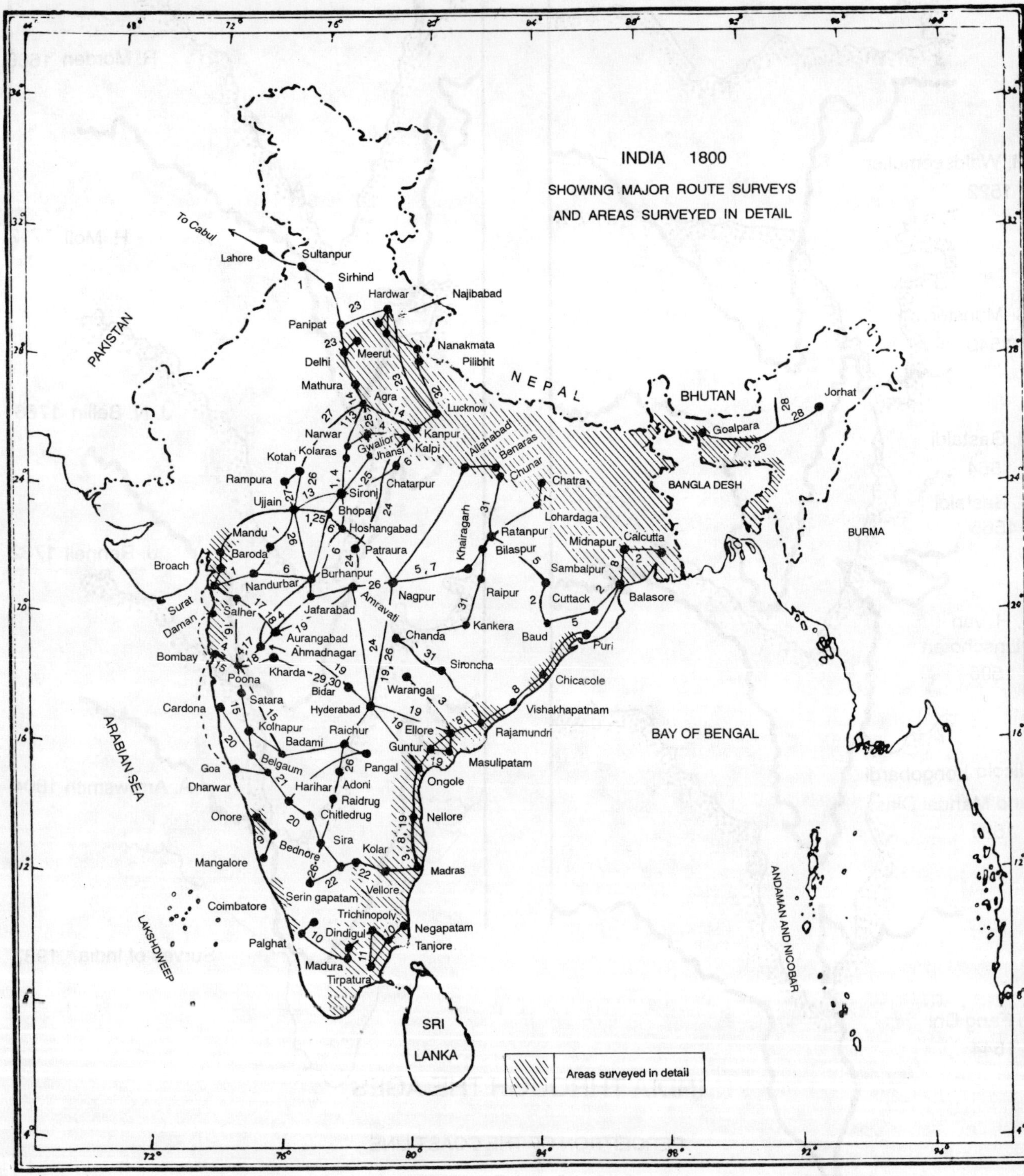

Map 22: *India, showing Major Route Surveys and Areas surveyed in detail* (AD 1800). (Compiled by the author from various sources).

Note: The outline of the map of India is the photographic enlargement of the Survey of India 1:21 million Plain Outline Map of India, printed 1987.

MAP NO. 22
REFERENCES

Sl.No.	Surveyor	Year of	From	To
1.	Father Monserrate	1579-81	Surat	Kabul
2.	Thomas Motle	1765	Calcutta	Sambalpur
3.	Robert Gardiner	1767	Madras	Warangal
4.	William Smith	1775	Kalpi	Bombay
5.	William Campbell	1778	Cuttack	Nagpur
6.	Arthur Caldwell and Duncan Stewart	1778-9	Kalpi	Surat
7.	James Ewart	1781	Chatra	Nagpur
8.	Colonel Pearse	1781-5	Midnapur	Madras
9.	Charles Reynolds	1782-3	Mangalore	Bednur
10.	Robert Kelly	1783	Negapatam	Coimbatore
11.	Robert Kelly	1783	Trichinopoly	Tanjore
12.	James Ewart	1784	Nagpur	Benaras
13.	Charles Reynolds	1785	Surat	Mathura
14.	Charles Reynolds	1785	Mathura	Kanpur
15.	Charles Reynolds	1786	Bombay, Badami	Poona
16.	Charles Reynolds	1787	Surat	Poona
17.	Charles Reynolds	1787	Poona	Surat
18.	Charles Reynolds	1788	Surat	Poona
19.	Charles Reynolds	1789	Madras	Surat
20.	G.L. Emmitt	1790-2	Cordona	Sira
21.	G.L. Emmitt	1790-2	Harihar	Goa
22.	R. Colebrooke and Alexander Beatson	1791-2	Madras	Seringa-patam
23.	Charles Reynolds and James Blunt	1792-3	Meerut	Lucknow
24.	Thomas Enburey and James Blunt	1792-3	Hyderabad	Kalpi
25.	Dr. Wm. Hunter	1792-3	Agra, Ujjain	Agra
26.	Charles Reynolds	1792-3	Seringapatam	Agra
27.	Dr. Wm. Hunter	1793	Ujjain	Agra
28.	Thomas Wood	1793-4	Assam	
29.	G.L. Emmitt	1795	Hyderabad	Poona
30.	Colin Mackenzie	1795	Hyderabad	Kharda
31.	James Blunt	1795	Chunar	Rajamundry
32.	Mark Wood	1799-1800	Kanpur	Najibabad

surveyors who had to move, generally, always a step ahead of the moving columns of the armies, to get information about the terrain the armies were supposed to traverse. It tells us of the efforts of the few who wanted the improvement of the geography of the country, of the trials of those who were interested in putting the worth of surveying on sound footing and make it grow as a serious profession in every respect. Maps, in those days, used to be drawn, copied and coloured by hand, with ever increasing number of topographical symbols invented for each of the varied items of vegetation, the types of terrain the surveyors came across and which they were never before familiar with. Beautifully hand coloured maps, with artistically drawn hechures, began to appear. These hechures had motifs and designs representing the artistic skill and the prevalent art forms of the time, which could very well be considered pieces of fine art. The maps had begun to be treated as works of art and as 'status symbols', fit for decorating the drawing rooms of the affluent in society. More than merely giving a written description, the maps gave a glimpse of the drama of unravelling the mysteries of the unknown and the romance and the thrill involved in the making of maps was in itself worth the experience. This was truly the age of romance with maps.

NOTES

1. R.H. Phillimore, *Historical Records of the Survey of India*, Dehra Dun, 1945, Vol. I, p. 209.
2. Susan Gole, *Early Maps of India*, New Delhi, 1976, p. 48.
3. Susan Gole, *A Series of Early Printed Maps of India in Facsimile*, New Delhi, 1984, Map No. 13.
4. Susan Gole, *India Within*, op. cit., p. 60.
5. Susan Gole, *Early Printed Maps*, op. cit., Map No.12.
6. Susan Gole, *Early Indian Maps*, op. cit., p. 54.
7. Ibid., p. 54.
8. Susan Gole, *India Within*, op. cit., p. 61.
9. Ibid., pp. 62-3.
10. Ibid., p. 60.
11. Ibid., p. 60.
12. Helen Wallis, 'Missionary Cartographers to China', in *Geographical Magazine*, London, 1975, Vol. 47, No. 12.
13. Ibid., p. 752.
14. Ibid., p. 754.
15. Andrew S. Cook, 'Maps from a Survey Archive: India Office Collection', in *The Map Collector*, The British Library, London, Map Collection, September 1984, No. 28, p. 27.
16. Susan Gole, *Early Printed Maps*, op. cit., Sl. No. 19.
17. Ibid., Map Nos. 19(a), 19(b), 19(c).
18. Ibid., Sl. No. 24.
19. Ibid., Map Nos. 20, 21.
20. Leo Bagrow, op. cit., p. 180.
21. Ibid., p. 179.
22. Public Archives of Canada, *Atlases Published in Netherlands*, op. cit., p. X.
23. Ibid., p. X.
24. Ibid., p. X.
25. Leo Bagrow, op. cit., p. 180.
26. Public Archives of Canada, *Atlases Published in Netherlands*, op. cit., p. X.
27. Ibid., p. XI.
28. Ibid., p. XI.
29. Leo Bagrow, op. cit., p. 181; Public Archives of Canada, *Atlases Published in Netherlands*, op. cit., p. XI.
30. Public Archives of Canada, *Atlases Published in Netherlands*, op. cit., p. XI.
31. Leo Bagrow, op. cit., p. XII.
32. *Public Archives of Canada, Atlases Published in Netherlands*, op. cit., p. XII.
33. Leo Bagrow, op. cit., p. 183.
34. Ibid., p. 184.
35. Susan Gole, *Early Printed Maps*., Sl. No. 27(a), 27(b); Map Nos. 27(a), 27(b).
36. Ibid., Map Nos. 29, 30.
37. Ibid., Map Nos. 25, 26.
38. Ibid., Map No. 30.
39. Ibid., Map No. 31.
40. Ibid., Map No. 32.
41. Ibid., Map Nos. 36, 37.
42. Ibid., Map Nos. 39, 42A, 42B.
43. Ibid., Map Nos. 34, 33a, 33b, 33c, 33f; Public Archives of Canada, *National Map Collection, French Atlases in the Rare Atlas Collection*, Provisional Series, No. II, Ottawa, 1974, Vol. I, p. 63, Sheet Nos. 26, 27.
44. R.H. Phillimore, op. cit., Vol. I, pp. 11, 148, 209.
45. S.N. Prasad, (ed), *Catalogue of the Historical Maps of the Survey of India, 1700-1900*, New Delhi, 1975, compiled in the National Archives of India, Map No. F.7/11.
46. James Rennell, *Memoir of a Map of Hindoostan*, 3rd edn., London, 1793, p. 23.
47. R.O. Cambridge, *An Account of the War in India between the English and the French on the Coast of Choromandel ... from 1750 to 1760*, London, 1761, map facing p. 15.
48. The first map was compiled in two sheets; first edition - 1873; second edition-1785, the second map was compiled in four sheets—first edition of the map, 1788; second edition, 1793; third edition, 1793.
49. Home Deptt., Public, Con., 30 May 1784.
50. R.H. Phillimore, *Historical Records*, op. cit., Vol. I, pp. 88, 141-2.
51. Survey of India, *Field-Books & Memoirs*, 1774-1866, in the NAI, No. 60.
52. Susan Gole, *India Within*, op. cit., p. 201, Map. No. 2.1.
53. James Rennell, *Memoir*, op. cit., 1793, pp. 280-1.
54. R.H. Phillimore, *Historical Records*, op. cit., Vol. I, p. 89.

55. James Rennell, *Memoir*, op. cit., 1793, p. 21; R.H. Phillimore, *Historical Records*, op. cit., Vol. I, p. 91.
56. R.H. Phillimore, *Historical Records*, op. cit., Vol. I, p. 92.
57. Ibid., p. 95. 118/12.
58. Amboor-Ambur.
59. James Rennell, *Memoir*, op. cit., 1793, p. 28.
60. S.N. Prasad, op. cit., Map No. 150/9.
61. SOI, *Field Books*, op. cit., No. 147.
62. C.R. Markham, *Memoir on the Indian Surveys*, London, 1878, p. 55.
63. R.H. Phillimore, *Historical Records*, op. cit., Vol. I, p. 4.
64. James Rennell, *Memoir*, op. cit., 1793, p. 28.
65. SOI, *Field-Books*, op. cit., Nos. 120, 134, 142, etc.
66. S.N. Prasad, op. cit., Map Nos. F.183/3, etc., 189/1, 2.
67. Ibid., No. F.118/12.
68. R.H. Phillimore, *Historical Records*, op. cit., Vol. I, p. 129.
69. General Deptt., Letters to Court, 12 Nov. 1761; A.S.Cook, op. cit., p. 7, n. 13.
70. James Rennell, *Memoir*, op. cit., 1793, p. 62; R.H. Phillimore, *Historical Records*, op. cit., Vol. I, p. 31.
71. Ibid., p. 62; ibid., p. 31.
72. SOI, *DDn. Vols., Correspondence Records of the Survey of India*, 1780-1900, N.A.I., No. 82/1800-10, p. 138, Letter from Sackville to the SG.
73. SOI, *Field Books*, op. cit., No. 272.
74. Ibid., No. 320, p. 27.
75. Ibid., No. 272.
76. S.N. Prasad, op. cit., Map Nos. F29/26-33.
77. Home Deptt., Public, Con., No. 19, 6 Feb., 1784.
78. James Rennell, *Memoir*, op. cit., 1793, p. 237.
79. S.N. Prasad, op. cit., Map No. 30/1.
80. R.H. Phillimore, *Historical Records*, op. cit., Vol. I, p. 42.
81. S.N. Prasad, op. cit., Map No. F.12/30.
82. Ibid., F.12/31.
83. James Rennell, *Memoir*, op. cit., 1793, p. 83.
84. R.H. Phillimore, *Historical Records*, op. cit., Vol. I, p. 68.
85. Ibid., Vol. I, p. 68.
86. Ibid., Vol. I, p. 222.
87. James Rennell, *Memoir*, op. cit., 1793, p. 253.
88. R.H. Phillimore, *Historical Records*, op. cit., Vol. I, p. 70.
89. Ibid., Vol. I, p. 74.
90. S.N. Prasad, op. cit., Map Nos. F.91/1-2.
91. R.H. Phillimore, *Historical Records*, op. cit., Vol. I, p. 80; Tirmizi, S.A.I. (Ed.), *Catalogue of the MRIO - Miscellaneous Maps of the Survey of India, 1742-1872,* compiled by the National Archives of India, New Delhi, 1982, Map Nos. 1781-1800/1-1A; Prasad S.N., op. cit., Map Nos. 171/21-2.
92. R.H. Phillimore, *Historical Records*, op. cit., Vol. I, p. 82.
93. Ibid., p. 14.
94. Home Deptt., Public, Letters to Court, 25 Jan. 1770.
95. R.H. Phillimore, *Historical Records*, op. cit., Vol. I, p. 17; India Office, *A Catalogue of Manuscript and Printed Reports, Field-Books, Memoirs, Maps, etc., of the Indian Surveys Deposited in the Map Room of the India Office ..., London,...* 1878, p. 548; F.C. Hirst, *The Surveys of Bengal by Major James Rennell, 1764-1777,* Calcutta, 1917, Sl. No. 127, p. 44; James Rennell, *Memoirs*, op. cit., 1793, p. 10.
96. R.H. Phillimore, *Historical Records*, op. cit., Vol. I, p. 101.
97. C.R. Low, *History of Indian Navy*, 2 vols., London, 1877, Vol. I, p. 16.
98. Ibid., Vol. I, pp. 185-6.
99. Ibid., Vol. I, p. 186.
100. India Office, *A Catalogue*; op. cit., 1878, pp. 104-5, 163, 548, 606-7.
101. F.C. Hirst, op. cit., 1917, pp. 43-7.
102. R.H. Phillimore, *Historical Records*, op. cit., Vol. I, p. 123.
103. Home Deptt., Public, Con., 22 Dec. 1783.
104. Ibid., Con., A., 9 Mar., 1778, No. 191
105. R.H. Phillimore, *Historical Records*, op. cit., Vol. I, p. 191.
106. Ibid., Vol. I, p. 102, n. 10.
107. C.R. Low, op. cit., Vol. I, p. 188.
108. S.N. Prasad, op. cit., Map Nos. F.137/51-3.
109. R.H. Phillimore, *Historical Records*, op. cit., Vol. I, p. 45.
110. Ibid., Vol. I, p. 45.
111. C.R. Low, op. cit., Vol. I, 186.
112. Foreign Deptt., Secret & Political Con., 22 Dec. 1788.
113. Ibid., Con., A, 3 June 1789; *Selections from the Records of the GOI*, Home Deptt., No. XXIV.
114. SOI, *D.Dn. Vols.*, op. cit., No. 10.
115. Selections from the Records of the GOI, Home Deptt., No. XXV; C.R. Low, op. cit., Vol. I, p. 186.
116. C.R. Low, op. cit., Vol. I, p. 186.
117. Ibid., p. 190.
118. R.H. Phillimore, *Historical Records*, op. cit., Vol. I, pp. 124-5.
119. C.R. Low, op. cit., Vol. I, p. 191.
120. Ibid., Vol. I, p. 191, footnote.
121. R.H. Phillimore, *Historical Records*, op. cit., Vol. I, p. 125.
122. C.R. Low, op. cit., Vol. I, p. 191.
123. Ibid., Vol. I, p. 191.
124. SOI, *Field-Books*, op. cit., Nos. 49, 227.
125. SOI, *D.Dn. Vols.* op. cit., No. 16, p. 83.
126. Ibid., No. 16, p. 135.
127. Home Deptt., Public, Letters to Court, 12 February 1771.
128. F.C. Hirst, op. cit., pp. I-II.
129. Ibid., p. 1.
130. Ibid., pp. 16-20.
131. Ibid., pp. 7-11.
132. Ibid., pp. 12-13.
133. Ibid., pp. 13-14.
134. General Deptt., Letters to Court, 12 Nov. 1761; A.S. Cook, op. cit., p. 7, No. 13.
135. S.N. Prasad, op. cit., Map No. F.51/6.
136. Ibid., F.52/5, 7-8 (copies).
137. R.H. Phillimore, *Historical Records*, op. cit., Vol. I, p. 92.
138. Ibid., Vol. I, p. 21.
139. S.N. Prasad, op. cit., Map Nos. F.140/1-6.
140. R.H. Phillimore, *Historical Records*, op. cit., Vol. I, p. 92.
141. Ibid., Vol. I, p. 92.
142. Ibid., Vol. I, pp. 92-3.
143. SOI, *D.Dn. Vols.* op. cit., No.1820-30/246, p. 142.
144. Home Deptt., Public, Co., 4 April 1764.
145. J.J.A. Campos, *History of the Portuguese in Bengal*, Calcutta, 1919, frontispiece Map.
146. R.H. Phillimore, *Historical Records*, op. cit., Vol. I, p. 221.
147. Imperial Library, *A Catalogue of Maps and Plans*, Calcutta, 1910, Sl. Nos. 411, 502.
148. Home Deptt., Public, Letters from Court, 16 March 1768.
149. R.H. Phillimore, *Historical Records*, op. cit., Vol. I, p. 222.
150. F.C. Hirst, op. cit., p. III.

151. John Malcolm, *Life of Robert, Lord Clive*, Murrey, 1836, Vol. II, p. 523.
152. Ibid., Vol. III, p. 133.
153. A.S. Cook, *Major James Rennell and A Bengal Atlas (1780 and 1781).* Foreign and Commonwealth Office, India Office Library and Records, Report, London, 1976, p. 9.
154. Ibid., p. 9.
155. R.H. Phillimore, *Historical Records*, op. cit., Vol. I, p. 223.
156. Ibid., p. 223.
157. Ibid., p. 223.
158. A.S. Cook, *James Rennell and Bengal Atlas*, op. cit., p. 11.
159. Ibid., p. 11.
160. India Office *A Catalogue.*, op. cit., 1878; A.S. Cook, *James Rennell and Bengal Atlas*, op. cit.; Prasad S.N., op. cit., Tirmizi, S.A.I., op. cit.
161. General Deptt., Letters to Court, 1 Jan. 1767; A.S. Cook, *James Rennell and Bengal Atlas*, op. cit., p. 9.
162. Ibid., 30 March 1767; A.S. Cook, *James Rennell and Bengal Atlas*, op. cit., p. 9.
163. Prasad, S.N., op. cit., Map Nos. F.146/2-3.
164. SOI, *D.Dn. Vols.*, op. cit., No. 1790-1800/12.
165. S.N. Prasad, op. cit., Map Nos. F.135/45, 150/7.
166. Ibid., F.135/45; 144/37.
167. Ibid., F.135/46; 144/37
168. R.H. Phillimore, *Historical Records*, op. cit., Vol. I, p. 131.
169. Ibid., p. 194.
170. Ibid., p. 7.
171. Ibid., pp. 285-6.
172. S.N. Prasad, op. cit., F.139/10, 13 (reduced copy).
173. Home Deptt. Public, Con., 8 Jan. 1767.
174. General Deptt. Letters to Court, 21 Nov. 1777. James Rennell had left India in April 1777.
175. R.H. Phillimore, *Historical Records*, op. cit., Vol. I, p. 37.
176. Home Deptt., Public, Con., No. 37, 13 Nov. 1783.
177. SOI, *Field-Books,* op. cit., No. 229; S.N. Prasad, op. cit., F.16/2.
178. Ibid., Nos. 206, 207.
179. Prasad, S.N., op. cit., F.56/4,5.
180. C.R. Markham, op. cit., 1878, p. 55.
181. S.N. Prasad, op. cit., F.150/8,9, 42-5; SOI, *Field-Books*, op. cit., Nos. 143, 144, 147, 148.
182. *Munshi* = an educated person.
183. *Harkara* = a messenger
184. Home Deptt., Public, Con., No. 21, 6 October 1783.
185. Ibid., Con., No. 53, 29 November 1784.
186. SOI, *Field-Books*, op. cit., No. 229.
187. Home Deptt., Public, Con., No. 2, 22 December 1783.
188. Ibid., Con., No. 21, 6 October 1783.
189. Military Deptt., Con., 10 March 1785.
190. Ibid., Con., 15 September 1785.
191. Ibid., Con., B, 7 Feburary 1786, No. 14.
192. R.H. Phillimore, *Historical Records*, op. cit., Vol. I, p. 216.
193. Home Deptt., Public, Con., 13 February 1786, No. 15.
194. SOI, *D.Dn. Vols.*, op. cit., No. 1790-1800/16, p. 9.
195. Home Deptt., Public, Con., No. 41, 22 October, 1788.
196. Ibid., Con., 13 October 1788.
197. Ibid., Con., 12 November 1788, No. 29.
198. C.R. Markham, op. cit., 1878, p. 56.
199. R.H. Phillimore, *Historical Records*, op. cit., Vol. I, p. 217, n. 1.
200. S.N. Prasad, op. cit., F.94/47, 95/30-53.
201. General Order dt. 29 Sept. 1788 issued by the Governor-General in Council.
202. Home Deptt., Public, Con., Dec. 5, 1794, No. 8.
203. Ibid., Con., 7 Sept. 1792, No. 40.
204. SOI, *D.Dn. Vols.* op. cit., No. 1810-20/146, p. 27.
205. Military Deptt., Letters from Court, 8 July 1795.
206. SOI, *D.Dn. Vols.* op. cit., No. 1810-20/146, pp. 5, 17.
207. R.H. Phillimore, *Historical Records*, op. cit., Vol. I, p. 219, n. 1.
208. S.N. Prasad, op. cit., F.95/28.

6. Modern Era

NINETEENTH CENTURY TO MID-TWENTIETH CENTURY

Till the beginning of the nineteenth century, the maps of India had been based mostly on route surveys, military or otherwise, with occasional but enough astronomical observations to link them with other maps, or to give them some sort of semblance of scientific surveys and accuracy. Though these are no longer considered scientifically accurate now, they served their purpose and served it well.

With the advancement in the science of map-making and the improvement of the instruments used therein, the cartographers discovered ways and means to prepare very accurate, almost perfect, topographical surveys based on triangulation, the system that gives a solid base to link all subsequent surveys with and to find and calculate any error, even in astronomical observations for latitude and longitude in such link-ups.

Though the system of triangulation was first used in France in the middle of the eighteenth century, it took about a century to master the art and improve the instruments to make sufficiently accurate maps.

In India, it all started with William Lambton in 1802. He took many years to make a network of triangles in the southern peninsula, and future surveys gradually began to be linked up with it. Meanwhile other types of surveys were in progress all around. Many had measured their own baselines. While on the one hand topographical, geographical and general maps were prepared for administrative purposes, efforts continued to be made to improve the quality of future maps.

We may now first discuss the triangulation of the country, before taking up other mapping activities in the post-eighteenth century period. The system of triangulation or the trigonometrical survey was generally considered to be the only accurate basis for the mapping of a country. In other words, to ascertain the geodelically accurate shape of the earth, or part thereof, e.g. a country, the measurement of an arc of the meridian is a must. The observations taken, therefore, during this

TABLE 3

Date	*Observers*	*Country*	*Latitude of middle of Arc*	*Arc measured*	*Length of degrees in feet*	*Length of degrees in Ordnance Table**
1735	Condamine and Bonguer	Peru	1° 31' 0" S	3° 7' 3"	362,808	362,758
1738	Maupertius re-examined by Svanberg	Lapland	66° 20' 10" N	1° 37' 19"	365,782	365,876
1738	Struve	Russia	58° 17' 37" N	3° 35' 5"	365,368	365,444
1750	Abbe' Lacaille	Cape of Good Hope	33° 18' 30" S	1° 13' 17"	363,713	363,871
1755	Bosovich	Rome	42° 59' 0" N	2° 9' 47"	364,262	364,479
1755	Mason and Dixon	U.S.A.	39° 12' 0" N	1° 28' 45"	363,786	364,235
1790	Delambre and Mechain	France	44° 51' 2" N	12° 22' 13"	364,535	364,599
1802	Roy and Kater	England	52° 35' 45" N	3° 57' 13"	364,971	365.097
1802	Lacaille	France	46° 52' 2" N	8° 20' 0"	364,872	364.730
1808	William Lambton	India	12° 32' 21" N	1° 34' 56"	363.013	362,929

*Geodetic Tables based on the elements of the figure of the earth given in the Account of the Principal Triangulation (of the Ordnance Survey), London, 1868. Quarto, 13 pages.[1]

activity by surveyors, were the 'forerunners of the Great Trigonometrical Surveys'. It was in 1735 that the first measurement of an arc of the meridian was undertaken in Peru in South America. This was done by the French expedition of Condamine and Bougner, aided by the Spanish brothers Ulloa. 'It consisted in the measurement of two bases connected by a series of triangles, one north and the other south of equator on the meridian of Quito, the arc being 180 miles long....'[2] Since then, meridian arcs in various countries, in different years, have been calculated. The measurements of arcs in various parts of the world, before it was calculated for the first time in India, are given in table 3.

Lambton was serving in India, in the army under General Harris who took part in the final war with Tipu. After the defeat and death of Tipu in 1799, Lord Wellesley was anxious to get accurate information respecting the newly acquired vast country. Dr. F. Buchanan was asked to report upon the agriculture and various products of Mysore and Malabar; Colin Mackenzie proceeded to topographically survey it. It was at that opportune moment that Lambton submitted his project or memorandum detailing his desire to measure an arc of the meridian and trigonometrically survey the entire peninsula. The memorandum was submitted in 1799:

> ... to Lord Clive the Governor of Madras, setting forth the advantage to general Scheme, and to the Geography of Southern India, which would arise from making it the scene of high Goddetic Operations, and in the first instance form a Geographical Survey across the Peninsula to conne Fort St. George with Mangalore, on the Malabar Coast—This proposition was immediately adopted, and an expensive establishment of Suitable Instruments was ordered from the first Makers in London—They arrived in 1802, previously to which Lieutt.Coll. Lambton had been engaged in Measuring a Base Line selecting great stations, and other preparatory Measures—[3]

William Lambton started his work in 1800 and the first baseline was measured near Bangalore. On 10 April 1802, on the arrival of new instruments, the actual commencement of the Great Trigonometrical Survey work began with the measurement of a fresh baseline near Madras, under the Madras government. The Surveyor-General, V. Blacker, wrote:

> In 1803-4 and 5 was completed the measurement across the Peninsula for fixing the positions of Several Meridian Lines, and determining the difference of Longitude between Fort St. George and Mangalore; and this was Succeeded by the Measurement at different times of an Arc on the Meridian extending Northwards from Punna[4] near Cape Comorin in Latitude 8°-9'-38"—.[5]

The measurement of the baseline was the backbone of the gigantic task not only of ascertaining the exact shape of the earth as a whole, but of finding out the correct shape of India. The process involved was—erection of a huge network consisting of a series of accurately measured triangles, the points of which were fixed by numerous astronomical observations, all of which were necessary, and were to act as springing-board from where, eventually, the triangulation was to be extended over the whole of India. The areas covered by these primary triangles were broken up by secondary triangles, similarly constructed inside the primary triangles. The smaller areas, were then filled in with detail from the topographical and revenue surveys and thus, the main features of the country were delineated within the fixed limits of errors.

The entire system of trigonometrical survey was divided into three clear-cut phases. The first phase involved the selection of sites for baselines which formed the ends of certain ranges of triangles. These were set out and measured with the utmost attainable accuracy. The baseline, AB, became a side of a triangle, the length of which was measured, and with the help of trigonometry, the distance of the third point, C, of the triangle which forms the third end of the baseline AB, could then be calculated. This third point C, was selected in such a way that it was visible from the other two points A and B, i.e. from the two ends of the baseline AB. The second part of the process was the construction of series of triangles. This was done by selecting another point, D, (the fourth point), which was visible from any two points of the first triangle. Then by calculating the sides of the second triangle, as one of its sides was a side of the first triangle also, the distance between the ends of this side and the fourth point, D, was calculated trigonometrically. By finding out the angles of the second triangle thus formed, the length of the sides of the second triangle could be ascertained. In this way erecting one triangle upon the other, a huge network of triangles could be created, the length of the sides of which could always be trigonometrically calculated. In actual practice, the network of these basic triangles was spread almost in a linear direction, either latitudinally or longitudinally.

The third phase of the survey was a further check on the accuracy of the triangles, and astronomical observations for latitude and longitude, made at selected points in the area so triangulated.

Thus, the bearings of a sufficient number of places spread all over the area of survey were fixed with the greatest of accuracy. These fixed points became the starting points for further secondary triangulation, and the points from which the topographical and revenue surveyors started their work of filling in the details, as well as of checking upon their accuracy.

To begin with the trigonometrical survey of a region, there must be an astronomically fixed point to build upon. For global calculations, the meridian of Greenwich Observatory at London is even now considered the Prime Meridian for observations of longitude, but for the trigonometrical survey being initiated in India in the early nineteenth century, the longitude of the Madras observatory was considered the primary longitude, and, therefore, its value was very carefully ascertained. This was obvious enough, as the entire structure of the survey network was to be built on this fixed point. If its value was incorrect, every station and place in that survey would be erroneous. 'In other words, the accuracy with which the entire map of India, as a whole, will be placed on the globe, will correspond precisely to the accuracy with which the geographical position of the Madras observatory has been determined....'[6] The baseline that was measured near Madras in 1802, was $7\frac{1}{2}$ miles long. Its measurement, which was started on 10 April, was completed on 22 May. It was after that the intricate process of the determination of the angle of the base with the meridian started.[7] With the completion of this began the process of the measurement of an arc of the meridian.

It may be remembered that a number of baselines were required at different places to countercheck the measurements and to minimise or eliminate errors.

From the Madras baseline, William Lambton carried the triangulation up to the Mysore plateau, and a second baseline was measured near Bangalore in 1804. This helped in the extension of the triangulation towards the Malabar coast. The earlier baseline near Bangalore, measured in 1800, was discarded. 'A series of triangles in 2° of latitude, was then carried across the peninsula; the flag staves at Tellicherry, Cannanore, and on Mount Delly, being intersected from the summit of Todiandamol, the highest peak in Coorg; but no baseline was measured on the Malabar Coast....'[8] The linking of the Malabar and Coromandel coasts in the peninsula through triangulation, was completed in 1806, linking the two sides of the peninsula. William Lambton then started thinking of the measurement of an arc of the meridian at Cape Comorin. For this, he had to bring his triangulation down to Cape Comorin. He measured a new baseline in Coimbatore in 1806. In 1808 another baseline was measured in Tanjore. From here Lambton proceeded eastward and carried his triangles to Negapatam. Then in 1809, he measured a baseline in Tinnevelly and the triangles were extended to the sea shore at Punnae, eight miles north-east of Cape Comorin. Another east-west line of triangulation was carried from Negapatam on the east coast to Ponnani and Calicut on the west coast.

In the meanwhile, the measurement of an arc in the peninsula of India was considered by scientists and the cartographers all over the world the work of the highest importance. This was done in 1808.

> By it the exact figure of the earth was to be ascertained; it should be remembered that this was not, as has been asserted, an object of mere curiosity. It affects some of the tables used in navigation, especially all those of which the moon's parallex is an element and is therefore an investigation of the greatest practical consequence to the whole civilized world. Moreover, the measured baseline, the series of accurately measured triangles, and points fixed by numerous astronomical observations all ... are necessary for ascertaining the shape of the earth....[9]

By 1813-14 the triangulation was extended from Cape Comorin to Goa on the west coast and to the mouth of the Kistna river on the east, comprehending almost all the territory in between.

> It may be supposed that with the accomplishment of this result, the most extended progress of the Geodetic Operations originally contemplated, may have been completed; but their importance was now felt, and they were accordingly allowed to proceed with every encouragement. In 1815, the Meridional series of Triangles was carried into the Nizam's Country as high as Bedar [Bidar], which besides adding another Section to the Arc prepared the foundation of the Survey of the Nizam's Dominions—The Triangulation was subsequently, extended to the Eastward as far as Cummamett [Khammamett] on the [south-east] Nizam's frontier, through all the tract comprised between the Rivers Godavery and Kistna, and the last work of the late Superintendent was the conduct of the Meridional Series to the vicinity of Ellichpoor, where it was closed by a base of verification. This is the fifth measured base which intersects the great Meridian of Doddagoontah, running thro' the Centre of the Peninsula, with an amplitude of no less than 13', completed by the labours of a single Astronomer.[10]

Soon after, at the end of 1818, George Everest joined William Lambton at Hyderabad. Lambton was now getting old. In spite of this, he measured a baseline at Takalkhera in the valley of Berar in 1822. He died on 20 January 1823.

> Col. Lambton, the first Superintendent of G.T. Survey, completed the triangulation of 165,342 sq. miles in the peninsula of India....[11]

By this time almost the entire peninsula had been opened up for topographical surveys and ready to be connected with the trigonometrical triangulation.

George Everest succeeded Colonel Lambton, and was appointed Superintendent of the Great Trigonometrical Survey of India in 1823. At this time Everest was engaged in the Bombay Longitudinal Series. On taking charge at Calcutta, he found that Lambton had measured a baseline at Takalkhera but the triangulation had not been extended further. Everest commenced the work in November 1823, carried the triangulation across Sultanpura hills to the plains of Sironj, about 15 miles north of Takalkhera. Here, another baseline was measured in November 1824. In January 1825, he took a series of observations at Kalianpur, near Sironj. But there he fell ill and was obliged to go on long leave to England to recover. He was still considered on duty though.

In England, Everest was forced to spend, by his illness, about five years. He spent these years in the study of all geodetic work then going on in Europe. He also kept himself busy in the design and construction of scientific apparatus that would be suitable in India. He was now fully equipped with the knowledge and mastery of the essential principles of geodesy then prevalent in Europe and was also acquainted with the latest instruments used therein. He came back to India in 1830, ready to cut out every possible source of error in the conduct of the geodetic su eys here.

While in England, through persistent efforts, he made the Court of Directors agree to the continuation of further extension of the triangulation and to accord priority to the extension of the meridional arc right up to the foot of the Himalayas from Central India, where the survey had reached at the time when he left India. This was a stupendous task. He also got the Court of Director's consent that the chains of triangles were to be observed and proceeded on a systematic plan and that the triangulation would ultimately cover the entire territories under the control of the Company and that of the friendly states. It was also decided by the Court that, as far as possible, no future topographical or geographical surveys be carried out until the framework of the main triangles was completed in the region.[12]

During his absence in England, his assistant in the trigonometrical surveys, Joseph Olliver, had extended a longitudinal series of triangles from Sironj base to Calcutta, about 700 miles away towards the east.[13]

On his return from England, the first task which Everest did, was to measure a baseline at Calcutta for verifying the line of triangles that was brought there by Olliver.

It is worth noting that while at England, Everest got manufactured, under his personal supervision, latest instruments, already being used in Europe for trigonometrical surveys—compensation bars for accurately measuring baselines; improved theodolites for triangulation; latest designed astronomical circles for arcs of amplitude, etc. For observations of trigonometrical stations from one point to another in the Indian weather conditions, he obtained heliotropes and reverberatory lamps for signals, etc. It was for the first time that compensation bars were used for the measurement of the Calcutta baseline, instead of the old and somewhat inaccurate method, by chains.

The Calcutta longitudinal series was brought from Kalianpur by Olliver. The Kalianpur Observatory was made the focal point of Everests' calculations and astronomical observations. This line was terminated at Calcutta, for which a termination and verification baseline was to be measured with compensation bars. The work on the baseline was commenced on 23 November 1831 and finished on 21 January 1832; but the final triangulation measurement was completed on 2 July 1832.[14]

In the meanwhile, on getting the report of Olliver's Calcutta longitudinal series, the Court of Directors had ordered that a number of chains of triangles should run northwards from this longitudinal chain from fixed points in such a way that the geography of the Bengal provinces could be covered and completed in a satisfactory manner by connecting important towns in the region.[15] To this, Everest had suggested, in brief, to run ten chains of triangles, about one degree apart, picking up important towns on the way, even if a little deviation from the meridian was necessitated. Everest outlined:

> I might mention Lucknow; Mr. Olliver's meridian of Khurara [Karara] would, when produced, nearly pass thro' this place, and a small branch might be thrown off so as to include Cawnpore [Kanpur], a point on which the map of the Lower Dooab mainly depends. Supposing any objection to exist to prosecuting operations of this nature in the kingdom of Oudh, the prolongation of the meridian of Umooah [Amua] would fall upon the latter place nearly. Allahabad, another important point, would be reached by the meridian of Goorwanee [Gurwani]; Furrukhabad and Bareilly would be reached by the meridian of Ranghir. With regard to more western points, they must follow the prolongation of the grand meridional arc from Sironj. . . . To the eastward, Mr. Olliver's series offers the Meridians of Gohrah [Gora] which prolonged would take up Azimgurh [Azamgarh] and Goruckpore [Gorakhpur], of Hooriloung [Hurilong] which would fix Patna and Bettiah, of Malooncha, [Maluncha] which would nearly fall upon Rajamahl and Purneah, and lastly of Parusnath, which prolonged in a southern direction would connect, and render available the map of part of that province....[16]

In short, this is a system which consisted of chains of triangles following particular obligatory meridians; at both ends, these chains were tied together by another chain which ran latitudinally following parallels of latitudes. The meridional chains were generally at intervals of about 1° apart and the longitudinal chains were drawn along the parallels of Calcutta, Bombay and Madras, a distance of about 6° apart. This system was called as 'Gridiron system' by Everest. It differed from the earlier system followed by Colonel Lambton in the southern peninsula. The latter system was also very useful for geographical and topographical purposes. It could be described, in brief, as the network of chains terminated at various points by chain-measured baselines. In the middle of the entire network of chains, a chain of principal triangles running north-south following a certain meridian was measured. This chain was very carefully measured, both the sides and the angles, as it was used for calculating latitudes and longitudes of all the fixed points.

The suggestions of George Everest were accepted by the Court, and he was asked to put forward detailed proposals as to the establishment, the cost involved and the time required.[17] Everest's reply was a very lengthy one, discussing the general policy of the triangulation and its extension, advantages of arranging the principal triangles in definite meridional and longitudinal chains, filling up of the subordinate chains by other survey parties, the areas to be covered, mode of operation in the flat country, use of towers and scaffolds, suggested constitution of eight parties—two for the Great Arc Series and six for other lesser series, and the cost involved including salaries and other expenses.[18] Long exchange of letters followed. Although the cost involved in the trigonometrical surveys was colossal many of the proposals of Everest were accepted.

As stated earlier, the baseline at Calcutta was measured between November 1831 and January 1832, and the triangulation of the Calcutta longitudinal series was completed in July 1832. Since then, because of long and tedious correspondence with the Government on his proposals for putting the Survey Department on a sound, scientific footing, Everest could not actively take to the fields.

It was in late December that Everest was able to resume his work connected with the Great Arc Series. The scenario of the operations was completely changed. Prior to this, the triangulation was being carried on in the plateau region of the peninsula where numerous mountain peaks were convenient to observe and mark as trigonometrical observation stations or points. But now, from the trigonometrical survey point of view the plains of northern India, beyond Sironj, were a very difficult country. Towards east from this point, the meridional series had to cross the flat land; on the north and north-west, the region was tolerable, as the country was marked by low hills. But from Delhi

> a wide plain, overgrown with groves of ... trees, and thickly scattered over with villages, extends for 104 miles to the foot of the Siwalik hills, the first outwork of the Himalayas. At a distance of a few miles from the observer on the great Gangetic plain, the trees appear to form a continuous belt of foliage, while clouds of dust often obscure the view.[19]

Everest's efforts with the available staff to start the triangulation northwards from Sironj became a frustrating job. This was compounded by the weather conditions. Due to poor visibility, he had to invent his own method of observing the trigonometrical stations, which could not be done by the prevalent methods so successfully used in the plateau region. His new system was called 'ray tracing system'. Another innovation of his was the substitution of heliotrope for day observations, which was used by Colonel Lambton, and reverberatory lamps with agrand burners, or blue lights, used at nights.[20]

'Ray tracing' could be described thus—owing to long distance between the stations to be observed, and the dusty weather in the plains region, the signals generally of blue light, were almost invisible. Since they were not visible, the direction in which to fix the telescope, was a problem. To obviate the delay or time-lag, he used to fix the telescope at a pre-determined point in a particular direction. A person was then asked to reach the fixed point by a series of minor triangles. This system, though cumbersome was the only system by which Everest could expedite the construction of chains of triangles in the plains. To move up from Sironj, between December 1833 to May 1834, he fixed 35 stations before he could reach Agra at the foot of the Siwalik hills.[21] This enabled him to measure the most northern base for the Great Arc Series which began by the end of 1834 at Dehra Dun and was completed by the end of March 1835. The observations on the Great Arc from Sironj to Dehra Dun baseline were concluded during the field season of 1836-7. On the closing of observations, the computation of the data collected from Dehra Dun towards Sironj was taken up. It was then that he came to know that the computed value of the baseline at Sironj differed from that of the measured one. This prevented the successful closing of the Great Arc Series. Everest then wrote to the Government about the discrepancy of which he was already apprehensive. He was sure that the measurement of the Sironj baseline that was initially measured with old and not so accurate instruments, would be the root cause of the non-alignment in the final link-up. He was also certain that unless the two bases were finally connected with each other by an intermediate series of triangles, no one could say with certainty as to the coincidence of the computed and measured baselines. He was positive that there was no error in the measurement of the Dehra Dun base, because in that case 'no latent source was to be dreaded'.[22]

Permission for re-measurement of the Sironj baseline was sought which was granted by the Government of India. In January 1838, A.S. Waugh carried out the re-measurement. Everest reported the satisfactory conclusion of 'the terrestrial work' between Sironj and Dehra Dun to the authorities.[23]

Knowing that the earlier observations in connection with the Great Trigonometrical Survey, undertaken by Lambton, were done with the help of outdated and inferior instruments leading to faulty conclusions, Everest tried to obtain permission for re-observing the entire triangulation between Sironj and Bidar. In this regard Everest had written to the Government of India in January 1838, emphasizing that though the baselines at Sironj and Takalkera now seemed to be in agreement, still other circumstances made it somewhat desirable that the triangulation should be revised. Only after this, he thought, would he be in a position to complete the triangulation between Bidar and Dehra Dun to his satisfaction by March or April of 1839. It was on the basis of the arguments put forward by him that he got the green signal to re-do the entire exercise.[24]

The Government agreed to this proposal reposing confidence in him and his judgement in the project—'in which large sums had already been expended'.[25] The re-measurement of the line from Bidar to Sironj was carried out during 1838-9. His request for a fresh measuring of the Bidar baseline was also approved because the old baseline, earlier measured by William Lambton, had disappeared and no sign or trace of the marking of the line was found on the ground. The Bidar baseline was measured by A.S. Waugh in 1841. On the other hand, the Government had turned down his request for re-doing Lambton's entire work right up to Cape Comorin. However, later on, Everest was proved right as the old stations, south of Bidar were found erroneous by Waugh when he tried to connect his surveys with them trigonometrically. Thus, during 1842 and 1843, Everest computed the data of his two main sections, Bidar to Sironj and Sironj to Dehra Dun.

In 1839, George Everest, realising that the main triangulation of the Great Arc Series was nearing completion and then some survey parties would be free and experienced hands might be lost to the Great Trigonometrical Surveys, he asked the Government for permission to make use of them in the subordinate meridional series, as and when these people were no longer required for the main arc. He was also aware of the closing errors of the Calcutta Longitudinal Series of Joseph Olliver as well. His request was passed on by the Government to the Court of Directors, who, initially, did not take note of it and replied evasively and pressed for the early completion of the triangulation, merely stating that such errors would not affect seriously the making of geographical maps. Thereupon, ensued a long and time-consuming correspondence between the Survey Department on the one side and the Central Government of India on the other. The interesting aspect of this was that the centre of controversy mostly rested on technical aspects of the entire survey operations, which the Government side on many occasions, misunderstood and misinterpreted, such as the use of the term 'Subordinate Meridional Series'. This was misinterpreted as 'minor series' or a series which was insignificant, thus degrading these surveys in importance. Everest very strongly protested the use of the term 'minor series' in place of 'Subordinate Series'. They perhaps, did not visualise that the 'Subordinate Meridional' surveys were equally important and needed to be equally accurately measured. Everest had to explain their utility and importance at great length.[26] Another point on which the controversy dragged on was that the object would not be served, as suggested by the Government, by running a longitudinal series direct from Agra or Delhi to the extreme eastern frontier and connecting this line with meridians coming up north from Calcutta longitudinal chain.[27]

Everest rejected the idea for a long longitudinal series, that too, running 'through a dead flat almost the whole way'. He explained that, being in the flat country, the progress would be very slow, and the side of triangles would not be more than 8 or 10 miles in length. This and other forceful arguments made the Government surrender. They allowed the longitudinal series emanating from the head of the Great Arc to connect the heads of Budhon, Ranghir and Amua series, but without infringing the Nepal territory.[28]

Seeing that the Great Arc Series was shortly coming to an end and the Subordinate Meridional Series would be earnestly taken up, the Surveyor-General had written to the Government, and in reply the authorities had posed certain questions to be answered in detail before any decision on the employment of the spare experienced hands on the Subordinate Meridional Series could be given

> First, The number, constitution, and strength of the parties to be formed for conducting meridional series.... Secondly,... The longitudes in which the parties shall respectively commence the new lines.... Thirdly, The manner in which you propose to test the accuracy of the work of each series upon its reaching the northern most limits.... Fourthly, If the system of meridional series at given distances is to be that adopted for Bengal, Behar, and Hindoostan, the Government will expect you to explain why the same plan should not be followed out to the southward —in the Nizam's territories and ... the Madras territory.[29]

Since Everest had in 1831 proposed ten parties for the ten subordinate series of Budhon, Ranghir, Amua, Karara, Gurwani, Gora, Hurilong, Chendwar, Parasnath, and Maluncha, he stood by it. These series would be approximately one degree apart. As the parties were not to cross the Nepal boundary, he abandoned his earlier proposal. Instead he suggested that after all the proposed meridional series had attained their northern-most limits, these would be joined by simple lines of triangulations to their west or east, as the case would be, till these simple series formed a continuous line running all along the Nepal border.[30]

Regarding the applicability of the same method of triangulation to the Nizam's territory, Everest stated that he did not get time to write to the Madras Government in respect of Hyderabad surveys or of Madras triangulation and promised to look into these matters.

While approving the programme of George Everest, the Government sanctioned only six parties instead of ten, and laid down the programme of work to be followed by these six parties. The Government also appreciated the 'gridiron' form of triangulation and approved of it as the best possible ground-work for a general map of India, but wanted to know the period within which the proposed consecutive meridional series would be completed. They also wanted to know if the Surveyor-General had any programme for that part of the country which still remained unsurveyed, namely the west of Delhi, and east of Calcutta meridional line of triangles and also for the unsurveyed tracts to the south of the Sironj longitudinal line.[31]

Giving the proposed period when the series would be completed, Everest answered:

... The series of Budhon, Ranghir, Amua, will be completed and connected with each other on the north side by June 1843. The series of Gurwani and Gora will in like manner be completed and connected with each other by June 1846. The other series to the east of Gora will each be similarly completed and connected with each other in four years from the time of its being taken in hand, which ought to be by June 1847.

This estimate will serve as a guide for ... plain countries which present by far the greatest difficulties....[32]

Because of the huge expenses already incurred on this project and because of what would further be required for the Great Trigonometrical Survey of the country, the Government while accepting the programme in general put certain restrictions

As the cost is a matter of importance, and as the number of meridians and their distances from one another is an essential element of the charge, the President in Council conceives that the object desired may be accomplished very nearly as effectually, ... and far more expeditiously, by argumenting the distance so as to diminish by one half the number of meridians between Calcutta and Seronj.

It is therefore desirable ... that a distance of two instead of one degree of longitude should be allowed between the lines of triangulation.[33]

The Government also categorically stated that the two meridional series of Gurwani and Hurilong were to be deleted from the programme.[34] Everest again insisted on the approval of his original plan and protested against the changes. His protest was referred to the Court of Directors by the Government. Their reply must have surprised the Government of India. Instead of condemning the expensive programme, they praised Everest. It was this approach of the Court of Directors at this crucial juncture that later enabled India to contribute with pride, her share of efforts in mapping the world on a uniform pattern in the committee of the international geodesists. The Court wrote

We unite with you in an anxious desire to carry on and finish at the least possible expense the trigonometrical operations which still remain to be completed. But our objects will not be effectually attained unless the work has a character of perfection and finality, which will render altogether unnecessary any future surveys.

We now learn that ... you have resolved that the meridians between Karara and Calcutta shall be diminished by one half, the distances between them averaging upwards of 100 miles. We apprehend that this ... will prevent that degree of accuracy which we are desirous to attain ... and will thus materially affect the permanent character of the work.[35]

The Court of Directors were apprehensive about the proposed programme of proceeding first with the meridians of Gora and Hazaribagh or Chendwar. Their objection was that the distances between Karara and Gora and between Gora and Hazaribagh, being respectively about 120 and 140 miles, there would arise the necessity hereafter for measuring intermediate meridians. They also ordered that in the meanwhile the progress of the Atlas of India in that direction be suspended.[36]

Earlier, in 1833 was taken up the Budhon Subordinate Meridional Series, which was the most westerly of the series, about one degree east of the Great Arc, and which following longitude 78°30' emanated from Joseph Olliver's Calcutta Longitudinal Series, north of Saugor. The line of triangulation passed through Jhansi, Gwalior and the Chambal valley and ran close to the Jumna river. It ran through districts of Agra, Mainpuri, Etah and Aligarh, crossing the Ganges, it reached the foot of the Himalayas and closed upon the stations of the Great Arc and the North-Eastern Longitudinal Series in 1842-3, after a period of about 10 years and covering about 400 miles.

Likewise, the second meridional series, Ranghir series, generally followed the meridian of 79°30'. It had a chain of single triangles. It crossed low hills, passed through Saugor, Jhansi, Hamirpur, Bundelkhand Agency, the plain of the Ganges river, entering Etawah, Farrukhabad and Bareilly, finally terminated in the great forests of the Tarai. The head of this series of meridional triangulation was linked and closed in 1842 by the Pilibhit Tarai Series, brought out by A.S. Waugh during 1842-3.

Amua Meridional Series was the third of the chain of meridional series from the west. Like other meridional series this also ran from Calcutta Longitudinal Series and generally followed 80°30'. It passed through the hills of Bundelkhand, Panna and Chhatarpur, through Banda, entered the valley of the river Jumna, and then Fatehpur and Kanpur, crossed the river Ganges, and passed through Lucknow and central Oudh. Its northern triangles finally merged with the North East Longitudinal Series; but before this could be done, it was temporarily connected with North Connecting Series in the Pilibhit Tarai area, pending the sanction to the North-East Longitudinal Series. The field operations of the Amua series were concluded in April 1840.

Finally, when George Everest left India at the end of 1843, the Great Trigonometrical Survey was on sound footing and under the stewardship of another able geodesist, A.S. Waugh. The Great Arc Series, Budhon, Ranghir and Amua series had been completed.

We have just seen, in a nutshell, the contribution of the great 'Father of Indian Geodesy', George Everest to the Indian Geodesy, who put it on a scientific basis that proved very useful in the twentieth century when international efforts to map the world on a uniform scale and pattern were initiated. A.S. Waugh succeeded him as Surveyor-General of India and was also appointed, like his predecessor, Superintendent, Great Trigonometrical Survey.

Waugh's first immediate charge was to complete the work of the triangulation left over by George Everest. At the time when A.S. Waugh assumed the responsibility of the Great Trigonometrical Survey, the position of the surveys was as follows.

Everest had projected gridiron[37] in northern India. The Great Arc and the Calcutta Meridional Series formed the two sides, and the Calcutta Longitudinal Series and the series running along the base of the Himalayas, being the other two sides. And according to his approved plan, ten meridional series, 60 miles apart, originated from the Calcutta Longitudinal line of triangles, ending in the Himalayan Series, forming the grating. Out of these ten meridional series, three series, Budhon, Ranghir and Amua, had been completed during the time of Everest.[38] As against this, the system, which William Lambton had adopted in the peninsular India, was to throw a network of triangles, supported on baselines, over the whole face of the country.[39] It fell upon Waugh to complete the remaining subordinate network of triangulation as envisaged in the plan.

It was during the time of Everest that Karara Meridional Series had started. It emanated from the Calcutta Longitudinal Series, from a hill called Karara and passed through Rewah, Allahabad and Lucknow to reach its northern-most limit in the Tarai. The work had started in February 1838[40] and was completed during Waugh's time in May 1845.[41] This series was next to Amua towards the east.

Next to Karara to the east, was Gurwani Meridional Series, but it was Calcutta Meridional Series which was taken up after when Karara Series was nearing completion in December 1843. This meridional series was the eastern side of Everest's gridiron and originated from the baseline measured during the time of Everest on the Barrackpoor road, at Calcutta, in 1831-2. The Calcutta Meridional Series was commenced in December 1843, passing north, parallel to the Hughly and Bhagirathi rivers and ended at a new baseline at Sonakhoda, near the foot of the Darjeeling mountains. The triangulation was completed in 1848.[42]

At the same time, when the Calcutta Meridional Series was started, i.e. in December 1843, Chendwar Meridional Series was taken up. This series was actually the eighth series to the east of the Great Arc. It passed through the Lower Provinces, and the Gangetic valley and reached its complete length in April 1846.[43]

The Gora Meridional Series, the sixth series east of the Great Arc, began its journey through Gorakhpur, the eastern part of the then North-Western Provinces, and across the Gangetic valley. The actual work was commenced in 1844,[44] although it had been constituted in October 1843.[45]

As in the case of Gora,[46] the Maluncha Meridional Series was also taken up in 1844. It passed, like Parasnath Meridional Series, north and south of the Calcutta Longitudinal Series. Passing through the Lower Provinces, it terminated in the year 1846.[47]

Next came the Gurwani Series which was taken up in December 1845. It passed through Jaunpur, Oudh and Faizabad and was completed in May 1847.[48]

The Hurilong Series was executed between 1848 and 1852. Passing through the Lower Provinces, it crossed the rivers Ganga and Gogra near their junction.[49]

Finally, the Parasnath Meridional Series was completed between 1850 and 1852. It also passed through the Lower Provinces.[50]

We have seen the ten meridional series growing out northwards of the Calcutta Longitudinal Series. The northern ends of all these series were connected with another longitudinal series called the North-Eastern Himalaya Series. This series was commenced in 1845 and was completed in 1850. It started eastwards from the Dehra Dun baseline to that of Sonakhoda base in Purnea.[51] This was a series from the principal trigonometrical stations of which all the major Himalayan peaks were visible and were fixed by measurements with the great theodolite. The Sonakhoda baseline was measured by A.S. Waugh in 1847-8, mainly to verify, both the triangulations of the North-Western Himalaya Series and the Calcutta Meridional Series. It was from this baseline at Sonakhoda, that, in future, another line of triangulation was extended eastwards into Assam.[52]

Thus, A.S. Waugh finished the work left to him by Everest. The gridiron which Everest visualised was completed. This gridiron of the trigonometrical series covered the whole of the north-west and the Lower Provinces, ready to be filled in by topographical and other surveys with their general triangulation network.

Simultaneously, with the progress of the gridiron surveys in the north, operations in the south were also progressing.

In 1842, the South Konkan Series was commenced from a side of the Bombay Longitudinal Series. It passed southward through Mahabaleswar to Goa, and was completed in 1844.[53]

The North Konkan and the Khanpisura series were then taken up. The Khanpisura Series extended as far as Neemuch on the Karachi Longitudinal Series. These series emanated from the Bombay Longitudinal Series, running east from Bombay to the Bidar baseline.

As in the case of the South Konkan Meridional Series on the western coast, the Madras Meridional Series on the east coast of the peninsula, which was the continuation of a series emanating from the Calcutta baseline, ran along the coast and went down to the Madras Observatory.

In the meanwhile, on the annexation of Sind in 1843, and the Punjab in 1849, Waugh drew a plan of preparing another gridiron, this time on the west of the Great Arc Series. His plan was based on the two baselines on the Great Arc, Dehra Dun and Sironj. From the northern base of Dehra Dun, a North-Western Himalaya Series of longitudinal chain was to be taken westwards to near Attock, and from Sironj base, the Calcutta Longitudinal Series was to be extended westward to Karachi; both these places are now in Pakistan. At both these stations, new baselines were to be measured. To have the fourth side of the line of triangulation, the Great Indus Series all along the Indus river, connecting Attock and Karachi, would be measured. Finally, a few intermediary meridional series, connecting the North-Western Himalaya Longitudinal Series and the westward extended Calcutta Longitudinal Series, would make the gridiron, as visualised by A.S. Waugh.[54]

Work on this plan started,[55] first on the North-Western Himalaya Series in 1847 from the Dehra Dun baseline. The line of triangulation reached Attock in 1853. The corresponding longitudinal line from Sironj was taken up at the end of 1848. This was also completed in 1853. The baseline in the northern longitudinal line was measured in 1851-2. This was at a place called Chuch Doab, near Attock. The other baseline near Karachi was measured in 1854-5. As already pointed out, the operations to connect these baselines by the line of triangulation, named the Great Indus Series, commenced simultaneously at both the ends. This series began in 1856 and was completed in 1860. The westward-extended Calcutta Longitudinal Series, extended to Karachi, was called the Great Longitudinal Series (Western Section).

The western most of the meridional series forming the bars of the gridiron of Waugh started southwards from near the city of Jhelum, (in Pakistan), but had to stop at Sutlej river, because the Great Desert beyond it could not be spanned with triangulation.

The next, the Gurhagarh Meridional Series, on the meridian of Amritsar, extended southward, passed close to Ferozpur on the eastern side and joined the Arumlia Series, which was brought up from near Neemuch on the Great Longitudinal Series (Western Section). This was laid down chiefly for Revenue Survey purposes at Ajmer in Rajasthan. It commenced in 1857 and was completed in 1862.

The eastern-most bar of Waugh's gridiron, the Rahun Meridional Series, took about six years to connect with the Great Longitudinal Series (Western Section), after starting from the North-West Himalaya Series. It was completed during 1857 and 1863.[56]

An oblique series, called the Sutlej Series, was brought up from Mitthankot (in Pakistan) on the Indus river near its junction with the river Sutlej, was connected with the southern end of the Jogitila Meridional Series and passed on further to terminate on the Gurhagarh Meridional Series. The series began in 1860 and was completed in 1863. This completed the gridiron of A.S. Waugh.

As it was forbidden by the Court of Directors to cross over to Nepal for trigonometrical survey purposes, the surveyors who were responsible for the survey of the following survey series, could observe and fix the position and height of the peaks of the Himalayan ranges, from east to west, from the trigonometrical stations of the following series:

1. Kashmir Series and triangulations along the Indus river.
2. North-West Himalaya Series
3. North-East Longitudinal Series and
4. Assam Longitudinal Series.

From the trigonometrical stations of the first series, the surveyors were able to observe the Karakoram range and the peaks in the Western Himalayas. As the name of the second series indicates, peaks in the Western Himalayas were visible from these stations. The North-East Longitudinal series stations were used for the Central Himalayan area and from the last series, the Assam Longitudinal Series, the surveyors could mark the peaks in the eastern Himalayas. During Waugh's period, another meridional series of triangulations were brought up from the Bombay Longitudinal Series to the Great Longitudinal Series (Western Section), near Neemuch, namely, the Khanpisura Meridional Series,[57] referred to earlier.

When J.T. Walker took over as the Superintendent of the Great Trigonometrical Survey, one of his first priorities was to measure a baseline on the east coast at a place which was nearly on the latitude of Bombay, because his plan was to extend the Bombay Longitudinal Series, which had already come up to Bidar, to the east coast. Vizagapatam (Vishakhapatnam) was the natural place which it could touch. By this time, the East Coast Series had already been completed from Calcutta to Vishakhapatnam. The measurement of the baseline was completed within two months, on fifth of November 1862. In December 1863, work on the connecting link of the East Coast Series, from Guntur to Madras, was taken up.[58]

As the Madras Observatory was the fixed point from which Colonel Lambton had initiated his triangulations in 1802, it was the nucleus or pivot on which the entire trigonometrical survey of the country rested. It was after 62 years that the triangulation was being brought back to the point from where it had started. In 1864, the East

Coast Series from Calcutta to Madras was completed.[59]

Earlier, Colonel Lambton, in his early surveys, had measured a chain of triangles from Madras Observatory to Bangalore. In 1865-6, a Madras Longitudinal Series was laid over on almost the same old ground from Madras Observatory to the Bangalore baseline for re-checking purposes. In the meanwhile, secondary triangulations were also completed from Guntur to Masulipatam, to fix the positions of the lighthouses on the coast. Points were fixed in the Ganjam district in 1882-3 for marine surveys.[60] During 1883-4 secondary triangulations between Puri and Balasore were also carried on for the erection of beacons, and the use of marine surveys.[61] In 1884-5, secondary triangles on the east coast from Vishakhapatnam to Coringa coast were completed.[62] Coconada was triangulated in 1885-6.[63]

From the Bangalore baseline, chains of triangulations were also carried to the west coast and to the extreme south to Cape Comorin during 1866-7. A new Bangalore baseline was ordered to be remeasured at a suitable place near Bangalore, for the same reason for which the Bidar and Sironj baselines had earlier been re-measured. The new Bangalore baseline was ready by 1868-9 and a longitudinal line of triangles was projected westward.[64] This was connected with the new series of triangulations, as well as with Colonel Lambton's Bangalore baseline which was measured in 1804 and with the astronomical station at Dadaguntha. Another new baseline was measured at the extreme end of the Great Arc, near Cape Comorin. This was done at Punnai, about 8 miles west of Cape Comorin, in 1868-9.[65] This was the last base, of the series of ten bases being newly measured, which were necessary for the verification of old triangles.[66] The operations were brought to a conclusion in 1879-80 by measuring the three polygons between Pondicherry and Madras.

In this way, the Bangalore Meridional Series was connected with Bidar baseline in March 1872 through its junction with the southern section of the Great Arc brought to the south by A.S. Waugh. The re-checking of the southern portion of the Great Arc was completed in 1874.[67]

During this year, another longitudinal series, known as the Ramnad Longitudinal Series, was completed which ran parallel to Ramnad at latitude 9°15' east of it. The intention was to connect it with the triangulations going on in Ceylon (Sri Lanka), which was done in 1876. It was ultimately connected with the principal triangulations near Madras.

In the eastern region of India to cover up the remaining part of the country, East Calcutta Longitudinal Series was extended eastward on the parallel of 23°N. This extension was meant for the surveys in the districts of Nadia, Jessore, Dacca, etc. (all in Bangladesh); and was completed during 1866-7. The topographical requirements of the then East Bengal (now Bangladesh) as well as Assam were amply met by one additional meridional series of triangles, running in the neighbourhood of the Brahmaputra river, though at a sufficient distance to secure the stations from the excursions of the river. This was called the Brahmaputra Series. It ran almost on 90° meridian. Commencing in 1867-8,[68] the series was completed in 1874.[69]

The Eastern Frontier Series began in 1861-2 from west of Gowahaty near the western end of the Assam valley. It ran down through Hill Tipperah, Chittagong (both in Bangladesh) to Akyab in Burma and further on to Thailand. During 1871-2 and 1875-6, triangulation of the Assam valley went on and reached eastern Sibsagar district in 1874; from Dibrugarh, where it had reached in 1875-6, it continued towards Naga Hills and also to the northeast across Brahmaputra to Sadiya. By 1877-8, the triangulation was extended up to Subansiri and Dihong valleys, in Arunachal Pradesh; and by 1866-7, on the other chain, it crossed into Burma.

There were some regions which were yet to be linked with the main triangulation of India. For this minor series were formed in those tracts which connected them with the major triangulation series.

In the south of the Bombay Presidency, a series of triangles known as the Mangalore Meridional Series, on the meridian of Mangalore, were commenced. They emanated from the Bombay Longitudinal Series, went down to connect, in 1873, with the Longitudinal Series which joined Mangalore with Madras.[70]

There were two more Meridional Series which emanated from the Calcutta Longitudinal Series, i.e. Sambalpur Series and Jabalpur Series. The first was started in 1864 and completed in 1866[71] and the second started in 1867-8 and was met with a line coming up from Bidar Longitudinal Series, to complete it. This was called the Jabalpur Meridional Series, described later.[72]

We have seen that J.T. Walker, on assuming charge as Superintendent of Great Trigonometrical Surveys, measured a baseline at Vishakhapatanam. By 1868-9, he completed the triangulation of the line between the Great Arc and the Jabalpur Series. This part of the series was called the Bidar Longitudinal Series. In 1869, it was further continued towards the coast, through the valley of the Godavari river. This was completed when it met the coast series in April 1872.[73]

Again, in 1867-8, triangulation work commenced on the median of 82°, emanating from the Bidar Longitudinal Series, going northward to meet the triangulation, being brought down from the Calcutta Meridional Series; the two met to complete the work in 1873. This was the Jubbulpoor (Jabalpur) Meridional Series.[74]

Other series, like the Eastern Sind Meridional Series, on the 70° meridian from the Karachi Longitudinal Series to the Great Indus Series, was in the region that now comprises Pakistan. This was commenced in 1876 and completed in 1881-2.[75]

To complete the triangulation of the north India, Kashmir Survey was commenced in 1862-3 and was completed by the end of 1864. The Kashmir Survey was a unique survey in which both trigonometrical survey and topographical survey were combined in the duties of a single party, under T.G. Montgomerie.[76]

Rajputara triangulation was the last to be taken up. In 1872, the Udaipur Meridional Series triangulation was commenced to complete the great northwest quadrilateral or A.S. Waugh's gridiron. This was closed in 1875 on the Sutlej Series in the north. It passed through Jaisalmer and Bikaner. After its completion, another series, called the Eastern Sind Series, was projected on 70° E from the Karachi Longitudinal Series to the Indus Series. This was started in 1876 at a side of the Karachi Longitudinal Series. It was mostly in the desert region and was completed in 1880-1, by closing on to a side of the Great Indus Series.

By about 1875, extensive chains of triangulatory measurements that were spread all over the country from the Himalayas to Cape Comorin were coming to a close as their main purpose was about to be achieved.

It is now time to recapture this vast venture, spread over a little more than half a century to have the idea of what it was and how it was achieved.

Between 1800 and 1825, a network of triangulation was being spread over southern India, based on and verified by several chain measured baselines. Through the middle of this network, a principal chain of triangles was carried in meridional direction, from Cape Comorin to Sironj in Central India on the meridian of 80°E. This chain was, later, carried further north to the Himalayas by George Everest and was called 'The Great Arc'. The angles of the Great Arc were measured with greater care and precision than those of the collateral network. At certain of its stations, astronomical observations for the latitude and for minor arcs of amplitude or azimuths were also taken up. Lambton died in 1823. During 1825-30, when George Everest, the successor to William Lambton, was in London on leave, he was able to procure latest equipments and instruments for survey purposes.

On his return in 1830, he, with the consent of the authorities in India and the Court of Directors in England, substituted the 'network system' of Lambton with his own system of triangulation, called the 'gridiron'. It consisted of meridional chains constructed at intervals of about one degree apart, and the longitudinal chains that would follow the parallels of Calcutta, Bombay and Madras, about five or six degrees apart. The whole was to be surrounded by external chains of the gridiron to follow the land frontiers of India and the coastlines in general. The entire triangulation was based on baselines measured with the Colby apparatus of compensation bars and microscopes to have a fixed standard of length. These baselines superseded the old bases of somewhat uncertain lengths. After George Everest, A.S. Waugh and J.T. Walker brought the great work initiated by William Lambton to a logical conclusion.

The huge data collected by the Great Trigonometrical Surveys was then to be reduced by computations. The stations of the principal triangulations—apart from other types of minor stations—in 1885, were 3,665 in number. These stations were constructed with a view to being as lasting and permanent as possible. The whole of the triangulation rested on the baselines measured with new instruments. The work of computation and reduction of the huge data began in 1869.

For the sake of convenience in the final reduction and computation, the chains of triangulations within the Indian borders, were grouped into five sections, the dividing lines of which were the two longitudinal chains of triangles, one connecting Karachi with Calcutta, known as the Great Longitudinal Series, and the other connecting Vishakhapatnam with Bombay, with the northern-most longitudinal line of chains running along the foot of the Himalayan range forming the north gridiron line. The central meridional chain of trigonometrical triangles running between 18°N Latitude and 30°N Latitude was called the Great Arc. Thus, for computation purposes, northern India was divided into four sections of quadrilateral shape explained above, with Sironj baseline at the centre. The four quadrilaterals were termed as North-East, North-West, South-East and South-West Quadrilaterals. In these sections, the main points had reference to the observations from the Kalianpur Observatory in Central India in post-1832 period. The fifth section, a triangular 'Trigon', south of the Vishakhapatnam-Bombay longitudinal line, with its calculations centred on the Madras Observatory, being taken as the point of origin of the southern triangulation by Colonel Lambton in 1802.

Simultaneously with the field operations, i.e. the measurements of the baselines and angles of the principal triangulations, the requirements of geodesy were not ignored. The astronomical observations for the determination of latitude and azimuth, and electro-telegraphic observations for the determination of differential longitudes were also going on at a number of the stations of the principal triangulation.

Another element, required in geodetic calculations, was pendulum observations at various stations for the investigation of variations of gravity and the figure of the earth. Later these observations were connected with the corresponding observations in other parts of the world.

Tidal observations were made to ascertain the mean sea-level and long lines of spirit levels too were carried out from the sea to the baselines of the principal triangulations.

All these various kinds of observations were carried

out in the service of Indian geodesy during the nineteenth century, which ultimately enhanced the prestige of Indian cartographers and geodesists in the world community.

In the meanwhile, in the mid-nineteenth century when J.T. Walker became the Surveyor-General, the hitherto three independent departments of Great Trigonometrical Survey, Topographical Survey and the Revenue Survey, were amalgamated into the Survey of India on 1 January 1878.

The designation of the Surveyor-General also changed to that of the 'Surveyor-General and Superintendent of the Trigonometrical and Topographical Branches'.

The authorities were now thinking of the reduction as computation of the huge and never data. The task was a stupendous one:

> the time had arrived for determining the procedure by which the fallible values of the several angles and baselines, as obtained by actual measurement on the ground, were to be rendered consistent, and final values were to be determined for the lengths and azimuths of the sides of the triangles and for the latitudes and longitudes of the stations, which is the ultimate object of all first class Triangulation .... The triangulation being formed of a large number of meridional chains tied together by a few longitudinal chains—forming sections somewhat resembling a gridiron in shape—presented a large number of circuits; and at the closing side of each circuit two values were forthcoming not only of the length of the side but also of its azimuth, and two values were also forthcoming of the latitudes and longitudes of the stations and its extremities. Thus three geodetic errors—as they have been called—had to be recognised and disposed of by a process of dispersion throughout the angles as well as the linear error....[77]

As stated earlier, the work of compulsion and reduction of the huge data was taken up in 1869 and the results were published in several volumes. Volume I, which contained the general account, was brought out in 1870. In 1878 and 1879, Volumes II, III, and IV were produced, giving a general account of the triangulation and its reduction, with full details of the North-West Quadrilateral; also Volume V, giving an account of the Pendulum Operations. This was followed in 1880 by Volumes VI for the South-East Quadrilateral; in 1882 by Volumes VII and VIII for the North-East Quadrilateral; in 1883 and 1887 by Volumes IX and X on the Electro-telegraphic longitude operations executed during 1875-7 and 1880-4, in 1890 by Volume XI on the Astronomical Observations for Latitude made during the period 1805 to 1885; and finally Volumes XII and XIII on the Southern Trigon. Of the synoptical volumes—which contained a precis of the results, both principal and secondary, for topographical and geographical requirements—twenty-three volumes have been published in all.[78]

It may be stated that along with principal triangulation a considerable amount of secondary triangulation was also completed. This was done by carrying out a number of chains to enable a link-up with the principal triangulation with important towns as envisaged by the authorities or because of the curiosity of linking up of prominent objects not connected directly with main triangulation but visible from principal stations, such as the snowy peaks of the Himalayan ranges, or because of the necessity of fixing points for the marine surveys, for the erection of lighthouses, or for linking up of certain towns or localities in the interior at a distance from the nearest principal chains, where, for topographical surveys, such data or points of chains of triangles were needed. However, much secondary triangulation remained to be done by the mid-nineteenth century, when J.T. Walker took charge of the operations.

During 1882-3, a chain of secondary triangulation was carried forward from the Mergni baseline, along the east coast mainly for the erection of beacons or lighthouses for the use of the navy or marine surveys. These operations were chiefly carried out in Orissa between Ichapur in Ganjam to the town of Puri. Chilka lake was included in this area. Measurements were taken at Puri and carried up towards Balasore during 1883-4. This opportunity was also taken to revise the measurements of the old stations erected in 1858-9. In the following season, 1884-5, a line of triangles was carried southward to link up with old triangulation of the Madras Coast Series which was terminated at Coringa, and further to the mouth of the Godaveri river. In 1885-6, triangulation was further advanced, both northwards and southwards, and by 1887-8 the entire coast from False Point to Point Calimere in the Tanjore district, was triangulated.

The series of secondary triangles emanating from the Great Indus Series and running along the parallel of 30°N latitude, in areas now in Pakistan, was also completed.

The operations on the Madras Coast Series, the Eastern Frontier Series, Eastern Sind Series, and in Assam Valley were also thus completed in the second half of the nineteenth century.

While reviewing the General Report of the Survey Operations during 1880-1, the Revenue and Agriculture Department (Surveys) of the Government of India observed—'The Principal triangulation of all India was brought to a close during the year under review by the completion of northern Section of the Eastern Sind Meridional Series....'[79]

We have reviewed the story of construction of the framework around which the edifice of the true map of India would be erected. The tale of the finding of details or geographical items with which the empty framework, with its numerous fixed points, was to be decorated to give it a shape, astronomically complete in every respect, which was called the accurate map of India, would now follow.

Side by side with the coming up of the framework, based on the Trigonometrical surveys, rough topographical details were initially determined mostly by

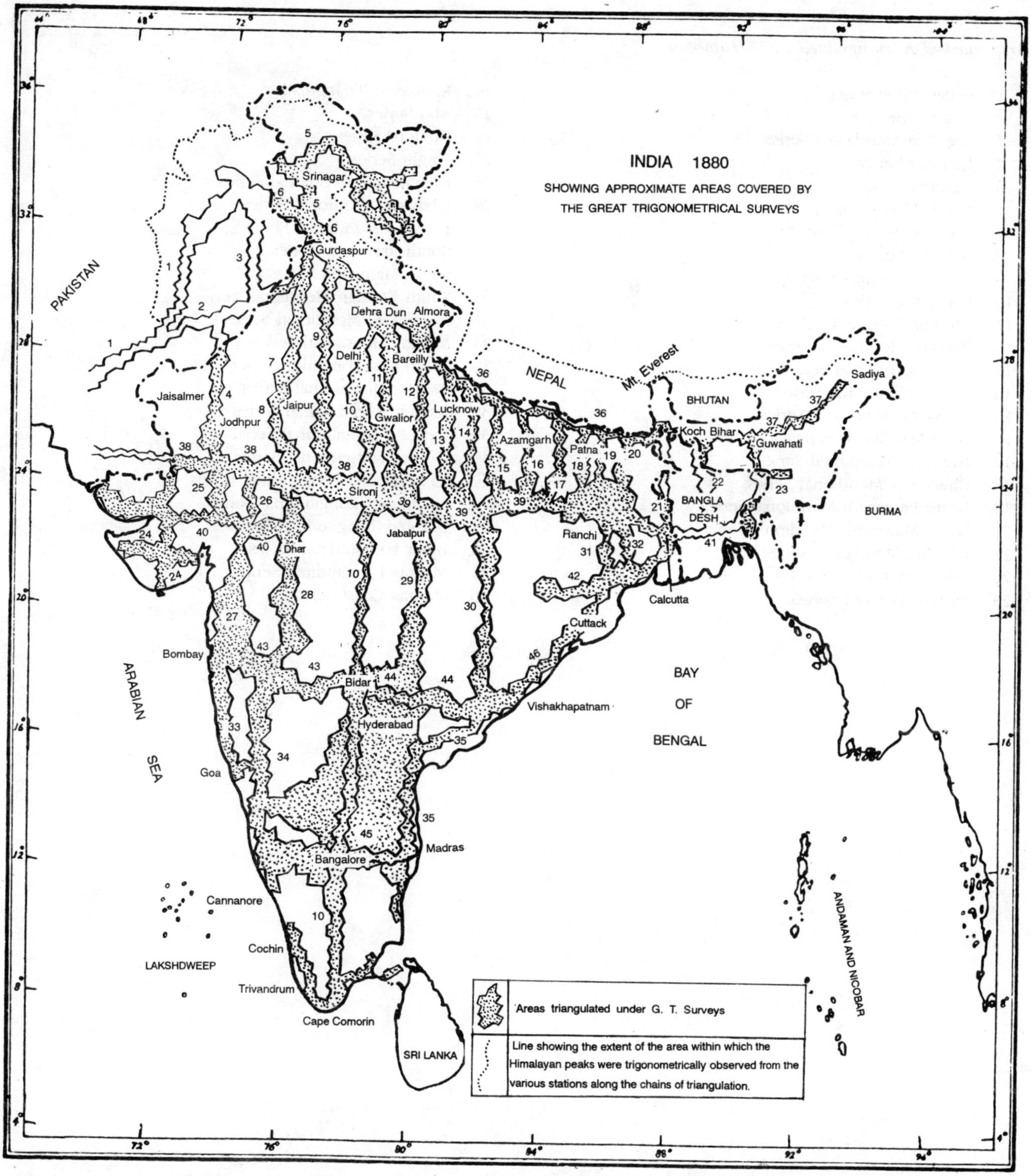

Map 23: *India showing areas covered by the Great Trigonometrical Surveys* (AD 1880). (Compiled by the author from various Reports of the Survey of India).

Note: The outline of the map of India is the photographic enlargement of the Survey of India 1:21 million Plain Outline Map of India, printed 1987.

MAP NO. 23
REFERENCES TO G.T.SURVEYS

Area marked A triangulated by W. Lambton

1. Great Indus Series.
2. Sutlej Series.
3. Jogi Tila Meridional Series.
4. Jodhpur Series.
5. Kashmir Series.
6. North-West Himalaya Series.
7. Garhagarh Meridional Series.
8. Arumlia Series.
9. Rahon Meridional Series.
10. Great Arc Series.
11. Budhon Meridional Series.
12. Ranghir Meridional Series.
13. Amua Meridional Series.
14. Karara Meridional Series.
15. Gurwani Meridional Series.
16. Gora Meridional Series.
17. Hurilong Meridional Series.
18. Chendwar Meridional Series.
19. North Parasnath Meridional Series.
20. North Maluncha Meridional Series.
21. Calcutta Meridional Series.
22. Brahmaputra Series.
23. Eastern Frontier Series.
24. Kathiawar Triangulation.
25. Abu Series.
26. Udaipur Series.
27. Singhi Series.
28. Khanpisure Meridional Series.
29. Jabalpur Meridional Series.
30. Bilaspur Meridional Series.
31. South Parasnath Series.
32. South Maluncha Series.
33. South Konkan Meridional Series.
34. Mangalore Meridional Series.
35. Madras Meridional Series.
36. North-East Longitudinal Series.
37. Assam Longitudinal Series.
38. Karachi Longitudinal Series.
39. Calcutta Longitudinal Series.
40. Gujarat Longitudinal Series.
41. East Calcutta Longitudinal Series.
42. Sambalpur Longitudinal Series.
43. Bombay Longitudinal Series.
44. Bidar Longitudinal Series.
45. Madras Longitudinal Series.
46. Madras Coast Series.

route-traverses of individuals, or military reconnaissances by surveyors attached with various detachments of the armies. These resulted in route maps with the help of which geographical maps of larger regions were compiled. Topographical maps were also attempted. Gradually, more accurate maps based on astronomically observed fixed points were prepared.

During the nineteenth century, Colin Mackenzie's contribution in the enhancement of the geographical knowledge of the country is noteworthy. He always insisted on regular surveys in an orderly system. It was his firm conviction that regular systematic surveys were the only ways of producing good maps. He was not only an indefatigable surveyor but a 'preserving collector of information' as well. His memoirs and journals, in original, preserved in the National Archives of India, New Delhi, State Archives of Tamil Nadu, at Madras, and at the India Office Library and Records, London, and elsewhere are among the 'most interesting relics', and proof of his intense interest in the field. C.R. Markham was of the opinion that '... Our knowledge of the literature and early history of Southern India is almost entirely due to the Mackenzie manuscripts.'[80] Commenting on the state of the incompleteness of the maps prepared under the orders of the Madras Presidency, Colin Mackenzie later attributed the failure to the wrong policy, pursued by the Survey Department and the Government, of frequently changing '... and removal of surveyors, & the want of a fixed establishment & office or Depot.... The employ of Engineers in one light promised fair at first, but the necessity of removing them for professional duties (in the army) occasioned greater inconveniences.'[81] It was the will, determination, and conviction of Colin Mackenzie that he could bring regular organization and system to the surveys of the Madras Presidency under whom he initially served in the early years of the nineteenth century, from 1800 to 1815. His draft regulations[82] issued for the guidance of the Survey Department at Madras in 1810 were approved by the authorities. His General Report[83] on the state of the Survey Department of Fort St. George in 1816-17, is ample proof of his sagacity and foresight in the matter. The sound organisation that Colin Mackenzie had given to the Madras surveys and followed by the close control maintained by his followers ensured that surveys of every district were satisfactorily concluded, though on way to their completion, the surveyors had often to change hands.

As mentioned in the last chapter, Colin Mackenzie, had surveyed a number of routes marched by the army against Mysore in 1790 to 1794. During the fourth Mysore war, he, along with his assistant Thomas Sydenham, kept a regular and complete survey of the route followed by the Nizam's army from Hyderabad to join the Grand Army. After the war was over in 1799, and Mysore shorn of the outlying districts under the Treaty of Seringapatam, some of the districts were given to Hyderabad as well. A Commission was set up to settle the details with the newly constituted government. Mackenzie, because of his reputation and deep knowledge of the Mysore country, was deputed to attend the Commission. It fell upon Colin Mackenzie to provide the best of maps that he could put together. The available material, on the outlying, remote districts of Mysore and on the boundaries of the state, was not accurate and sufficient enough and was sometimes conflicting. He himself said:

> ... it appeared absolutely necessary to commence such a work from an entire [ly] new foundation, for ... though the central part of the country had been repeatedly traversed by our armies in the Mysorean wars, yet the limits and extent of the several Districts were not defined, nor were even any plans of the surveys ... to be procured at this time, unless we except [accept] the results published so far back as 1792 by Major Rennell, and some manuscripts of detached parts in private hands.[84]

In spite of this state of affairs and the kind of survey material available at that time, Mackenzie manage to provide some useful maps. It was, therefore, decided that a survey of Mysore and the newly ceded districts, should be taken up immediately so as to put the matter under control. The Governor-General, even before he left for Calcutta, at once ordered Mackenzie to take up the task. Mackenzie set out on the job in 1800, along with an Assistant, John Mather. He worked for some years and by 1815 almost the entire work entrusted to him, was completed. The ceded districts were surveyed by the native surveyors of Mackenzie, under his direction, in 1809 to 1815. These were based on his own measured independent baselines and triangles.[85]

Colin Mackenzie was known for his accurate topographical maps based on his own system of triangulation. During that very time William Lambton was also engaged in the work of triangulation in the south, but Mackenzie worked on independent of Lambton. The latter measured his own baselines for the purpose in convenient places, each baseline from three to five miles long and connected them with triangles. He was able to make one general and seven provincial maps. His valuable memoir describing the maps was in seven folio volumes, containing narrative of the survey together with statistical, historical and antiquarian information.[86]

In 1809, Mackenzie became the Surveyor-General of Madras but during 1811-14, he had to leave India on an expedition to Java. On his return in 1815, he continued supervising the survey of the Ceded Districts which had begun in 1809, until he went to Calcutta and became the Surveyor-General of India (1815-21).

James Garling, one of the brilliant pupils of the Madras Military Institution, was able to survey Goa. He measured a baseline for the purpose near Cape Ramas in November 1810. He was busy with the survey of Goa from 1811, till

January 1813. His maps[87] of Goa are a feast for the eyes. These maps, along with their memoirs, are preserved in the National Archives of India at New Delhi.[88] When the survey of Goa was completed, James Garling, along with P.F. Conner and others began to survey Soanda and Bilgi in the North Canara in 1813. For this survey, Garling found two baselines already measured, quite handy, one from the Goa survey and the other a side of a triangle of the trigonometrical survey station measured by William Lambton. This survey was carried out on quite a detailed plan, on the scale of 1-inch to a mile, and was completed in 1815.[89] Garling's memoirs,[90] in six volumes, containing the history of the region, general description with details regarding its revenue, agriculture, trade and commerce including cultivation, water supply or irrigation, inhabitants, tenure and routes, and triangulation data, are in the custody of the National Archives of India. The volumes cover the period from 1814 to 1815.

P.F. Conner was then asked to survey Coorg. He did not have much difficulty in carrying it out. Since William Lambton had already triangulated it, Conner found a readymade basis for his survey which began in 1815 and was completed by October, 1817.[91] Maps of Coorg[92] by Conner, on four manuscript sheets, covering the period 1815-17 along with a reduced map and four volumes of his 'Memoirs'[93] are with the National Archives of India. The topographical details were added in the maps by plane-table survey. The Coorg Survey was considered a very valuable and complete survey.[94]

By 1844 almost all the districts under the Madras Presidency were topographically surveyed and mapped. Salem district was the last to be mapped and the survey party was then disbanded in 1844.

Tinnevelli district was surveyed by Thomas Turnbull, T. Hill, and others during 1807 and 1813. Though based on triangulation and measured baselines, it was not considered 'valuable' by the later authorities.[95] Its memoir, by T. Turnbull, gives, apart from the details of the surveys, its ancient history, translations of the inscriptions on granite walls of temples and other monuments, climate, population, produce, its trade and commerce and routes, etc.

The survey of the Travancore and Cochin states was commenced in 1816 and concluded in 1820-1. B.S. Ward and P.F. Conner, with numerous assistants, surveyed the districts. It was based on William Lambton's triangulation and only minor triangles were carried out to fill the interior details by plane-table. The survey was executed in an admirable manner and was considered an excellent geographical, topographical and statistical survey.[96]

The Madura district and the Manapara Polliams were surveyed between the years 1803 and 1806. As these were found to be very imperfect, re-survey of those districts was ordered on 22 February 1825.[97]

Ramnad and Poodoocottah were surveyed in 1871 to 1874. The survey was based on Lambton's triangulation.[98]

The survey of Dindigul was commenced in 1815 and was completed in 1816. This was done by Assistant Surveyors T. Turnbull and others. It was later considered inadequate and was re-surveyed by B.S. Ward and others during 1821. They also completed the Dindigul-Warsenaad valleys and the mountains of Anjinaad. The survey was also based on Lambton's triangulation.[99]

Trichinopoly district was first surveyed between the years 1803 and 1810 by Assistant Revenue Surveyors. Though it had independently measured baselines and triangulations, later connected by Lambton's triangulation, but, because it was executed in a very loose style, many complaints of great inaccuracy were received by the Survey Department from the Collector of the district. Re-survey, therefore, of the Trichinopoly district was ordered on 22 February 1825.[100] Ward could take up the survey only in 1832 after finishing the survey of Madura district. He could complete the fieldwork of the survey of Trichinopoly by April 1834. He went back to the army on promotion in November 1834. The maps together with the field-books and Ward's memoirs could be submitted to the Surveyor-General in May 1835.[101]

Part of the Allambaddy *paragana*, belonging to the Coimbatore district, was surveyed in 1807 by the surveyors engaged in the survey of Mysore. But the area was re-surveyed in 1809 when it was based on Lambton's triangulations, and was completed in 1812.[102]

Nilgiri mountains were surveyed in 1822, and south Coimbatore was surveyed in 1823. These also, like others of the region, were based on the William Lambton's surveys.[103]

Salem district, though based on triangles, was not connected with trigonometrical triangulations. Topographical details were very inaccurately depicted.[104] B.S. Ward was asked to re-survey it in 1825, but because of his other priorities, C.T. Hill took charge from Ward and started the survey in March 1835. The surveyors changed frequently—from Hill, the responsibility went over to John Campbell; from John Cambell, it fell on C.R. McMahon and again to Campbell who finally reported the completion of the fieldwork of the survey of Salem district in 1838.[105]

The districts of Pondella, Durshee, Pelloor of the Raja of Venkatagherry of the Nellore region, were surveyed in 1813 and the *jagir* district, Udaigiri and Pananoor, were surveyed in 1814.[106]

The Guntur district survey was started in 1815 by the Officers of the Military Institution, Madras, who worked till 1816. In 1817, it was Francis Mountford, who was ordered to survey it for revenue purposes as opposed to the earlier objective, being military or topographical one.[107]

Districts Ellore and Rajahmundry, were surveyed between 1815 and 1824. District Rajamahendri was surveyed on the basis of baselines measured near Kolar lake, independently of William Lambton's baselines, but the triangulation was connected with two fixed points of Lambton's triangles near the Kistna river.[108] The survey was probably commenced in 1821 and was completed in 1824.[109]

The districts or Circars of Condapilly, Masulipatam and Ellore formed the Masulipatam Collectorate. The survey of Condapilly, Masulipatam was commenced in 1816. It was based on the triangulation of the Great Trigonometrical Surveys and was completed in 1818.[110]

The survey of Ellore was based on independently measured baselines and triangulation. It was started in 1819 and was completed in 1823.[111]

In 1815, the survey of the Southern Circars began from the Kistna river northwards and by 1828, the survey party entered Ganjam, the survey being completed in 1833. The maps were compiled and submitted by 1844.[112] The survey operations in Nellore commenced in 1832-3 and completed in 1842. The final papers, like report, field-books, memoirs and the maps, etc., were submitted after completion, likewise, on 13 January 1844.[113] The Surveyor-General, A.S. Waugh, had favourably commented upon the Ganjam surveys in 1860.[114]

The Hyderabad state came under regular survey in 1818, by James Garling. Before the survey was completed, the work was passed on from James Garling to F. Young, then to S.C. Macpherson in 1829, who took leave in 1832, handing over charge to J.M. Du Vernet, returning only in July 1836. In the meanwhile, in 1831, the survey responsibility was passed on to the immediate control of the Supreme Government—to the Surveyor-General, by the Madras Government. The survey work suffered considerably because members of the surveyor party were ill during most of the 1830s, mostly because of jungle fever that threatened one member of the party after another.[115] To expedite the work, more hands were demanded from the authorities, but none was provided. By 1840, the progress of the survey was as follows.

From Raichur, where the survey work began in 1816, the whole of the *Doab* area between the rivers Tungabhadra and Kistna, was completed by 1819; then the survey work proceeded across Kistna river and continued through the districts of Pangul, Devarkonda, Nalgoonda, Hyderabad, and Khammammett and completed in 1824; Nilkonda and Gulbarga were completed in 1825; Medak was surveyed in 1828; Warangal and Kowlass came under survey in 1831; Bidar, Kalyani, and Naldrug in 1834, and Osmanabad in 1835.[116]

Thus, we have seen that the surveys under the Madras Government, progressed well and almost all the districts were topographically surveyed; wherever this was not so, or the work was being delayed, it was because Lambton's triangulation had not reached the district, and, therefore, the surveyor had to initiate his own triangulation by measuring special baselines and connecting them with his own triangles, as in the case of Nander and Pathri *Circars*, part of Aurangabad districts, etc.

On the western side of the peninsula, in the Bombay Presidency, Goa, North Canara, Malabar, Gujarat, Cutch, and Kathiawar were topographically surveyed between 1813 and 1820. Some maps from route surveys of the Deccan were compiled; Malwa was geographically surveyed between 1812 and 1816.

Early revenue surveys of Poona and Ahmadnagar districts were not useful for topographical information. Same was the case with maps of certain areas of Gujarat as well. The idea was for the first time mooted in 1840 that the surveys be rendered subservient to geography and topography.[117] John Pringle's surveys of the districts of Poona and Sholapur were, therefore, revised by 1841. By 1833, the surveys of southern Konkan were completed. Thus, we see that not much had been topographically surveyed in the Bombay Presidency by 1834, the period which saw a vast change in the policy of the Supreme Government so far as the Revenue and Topographical surveys were concerned.

But before 1834, in northern India, W.S. Webb was continuing with the general surveys of Kumaon from 1815 to 1820. Garhwal was surveyed in 1818. The Bundelkhand survey was undertaken by James Franklin in 1815 and completed in 1821. In Central India, especially the Bhopal district was surveyed in 1821, based on route surveys. Surveys in Azamgarh and Jaunpur districts were taken up in 1818. The districts of Panipat, Bhatinda, Delhi, Rohtak, Gurgaon, Mathura and Agra were surveyed between 1822 and 1842. Though surveyed for revenue purposes, the areas were surveyed on scientific principles and thus the topographical features were also properly laid down. The districts around Benaras were similarly surveyed during 1839 and 1841.

In Bengal, Sundarbans were re-surveyed between 1812 and 1818; and in the Lower Provinces, the Cuttack surveys were initiated in 1818.

One of the many epoch-making decisions of the Supreme Government was the initiation of Revenue Surveys in the Upper Provinces in 1822. J.A. Hodgson was asked to take up the revenue survey of the Upper Provinces or the North-Western Provinces. He believed in high standards of professional work and accurate survey and delineation of topographical features. This aspect of the survey, namely, scientific and accurate surveys with proper topographical features, was actually of no proper use for an accurate assessment of revenue. But it caused delay in the settlement process, and this, the administration did not want.

We are interested in examining revenue maps here not for the purpose of land revenue settlement but for their use as aids to topographical mapping. It was accepted by all concerned in early days that, since land revenue was a very large part of Indian revenues, the only system by which land settlement could be fair to the cultivators, *zamindars*, and Government, must be based on accurate professional surveys. Without knowing what proportion of land was productive and what was lying waste, the assessment could not be fair and this was possible only by scientific surveys, which were time-consuming. But, on the other hand, such maps were helpful for the compilation of topographical maps, as well.

Lord William Bentinck, the Governor-General, had taken much interest in revenue settlement and in maps and surveys, and was of the opinion that much more output could be got from the existing survey operations. The activities must be more closely associated with the requirements of the land settlements. To study the problem more closely, he circulated a statement of problems to all concerned. A number of replies and reports were received, explaining in detail the short-term and long-term requirements and benefits, mostly of the existing practices. A conference of the surveyors was held at Allahabad on 21-3 January 1833, presided over by the Governor-General, Lord William Bentinck himself.

Till then, topographical details in revenue survey maps were tolerably well executed, until the decisions at the conference altered the situation. The single object of the conference was to speed up the preparation of such maps that could be used in order to augment the system of revenue settlement. The whole exercise was a great disappointment to professional surveyors:

> A new plan was therefore adopted, introducing economy and rapidity, and sacrificing quality for quantity. The maps were only required to delineate village boundaries and sites with rough outlines of roads and the course of rivers, and were mere skeleton sketches. They preceded the Great Trigonometrical Survey, and thus proper connection was not in the first instance established between the two operations.[118]

Because of the first Afghan war between 1839 and 1842, the British Government to economise in the expenditure ordered retrenchment in every department[119] in 1842. Progress of the revenue surveys, both in the Lower Provinces and the North West Provinces during the period, is reflected in the table of surveys detailed hereafter. Almost all the surveys were either closed down or were expedited so that these might be completed by 1842, the date announced, by which all the surveys with very few exceptions, were to close down.

Before the retrenchment orders came into force, the progress of the revenue surveys of the districts of the Lower Provinces, until 1843, being surveyed by professional surveyors under the control of Sadr Board of Revenue, Lower Provinces, with their Headquarters at Fort William, Calcutta, is given in Table 4.

The districts mentioned in Table 5 constituted Revenue Divisions in the British administered region, in the 1840s in Northern India.[120]

TABLE 4

Sl. No.	*District*	*Taken over or Commenced in*	*Completed*
1.	Sundarbans	Jan. 1828	1835
2.	Lower Assam	Nov. 1832	1842
3.	Monghyr (Munger)	Dec. 1835	1837
4.	Tippera	Nov. 1836	1842
5.	Monghyr (Munger) and Bhagalpur	Jan. 1837	1840
6.	Jaintia, Sylhet and Cachar	Jan. 1837	1842
7.	Cuttack	Dec. 1837	1842
8.	Puri	Jan. 1838	1841
9.	Balasore	Mar. 1838	1842
10.	Midnapore	Apr. 1838	1844
11.	Assam, Upper	Nov. 1838	1842
12.	Hijli	Jan. 1839	1841
13.	Assam, Mattack Tea Area	Feb. 1841	1842
14.	Patna	Dec. 1841	1843
15.	Gaya	Jan. 1842	1843

It was after the completion of the district surveys that satisfactory maps of the respective Divisions could be compiled. The earlier district maps that were compiled from the revenue surveys generally had astronomical control.

Almost simultaneously with the orders of the revenue surveys in the North-Western Provinces and the Lower Provinces issued in 1822, the Court of Directors issued in 1823 another set of orders which had far-reaching effects on the general mapping activities in India, regarding the layout, projection, compilation and engraving of the proposed standard quarter-inch Atlas of India. This project would be discussed later.

The early revenue surveys of the North Western Provinces until 1834 were based on strict scientific topographical surveys, and were considered valuable maps for topographical purposes; but after 1834, in many cases, these were not considered proper for the Indian Atlas sheets. At that time all maps were being examined with reference to their suitability or otherwise of the Atlas.

The then Surveyor-General of India, George Everest, and his Deputy, James Bedford, were aware of the fact that a topographical map, based on accurate and scientific principles, must be connected or linked up by the Great Trigonometrical surveys to qualify as a correct scientific topographical map and would be accepted for

TABLE 5

Sl.No.	*Name of Division*	*Sadr or H.Q. Station*	*Component Districts or Zillas*
1.	Delhi Territory	Delhi	Delhi, Panipat, Gurgaon, Hissar, Rohtak.
2.	Saugor & Narabada Territories	Jabbulpore	Jubbulpore, Saugor, Seoni, Hatta, Rohlie, Hoshangabad, Narsingpur, Baitul.
3.	Meerut	Meerut	Meerut, Saharanpur, Muzaffarnagar, Bulandshahr, Dehra Dun.
4.	Agra	Agra	Agra, Aligarh, Mathura.
5.	Farrukhabad	Farrukhabad	Farrukhabad, Mainpuri, Sirpurah (Etah), Etawah, Helah.
6.	Moradabad	Moradabad	Moradabad, Nagina (Bijnour), Saheswan (Budaun).
7.	Bareilly	Bareilly	Bareilly, Shahjahanpur, Pilibhit, Kumaon.
8.	Allahabad	Allahabad	Allahabad, Fatehpur, Cawnpore (Kanpur).
9.	Hamirpur (Kalpi)	Hamirpur (Kalpi)	Hamirpur (Kalpi), Banda (North & South Bundelkhand).
10.	Benares(Varanasi)	Benares (Varansi)	Benares (Varanasi), Mirzapur, Jaunpur.
11.	Gorakhpur	Gorakhpur	Gorakhpur, Azimgarh, Ghazipur
12.	Sarun	Sarun	Sarun, Shahabad, Tirhoot.
13.	Patna	Patna	Patna, Behar (Gaya), Ramgarh (Hazaribagh)
14.	Monghyr (Munger)	Monghyr (Munger)	Bhagalpur, Monghyr (Munger), Malda, Purnea.
15.	Baulea	Baulea	Dinajepore, Rangpore, Rajshahi, Bogra.
16.	Murshidabad	Murshidabad	Murshidabad, Birbhum, Naddia.
17.	Dacca	Dacca	Dacca, Dacca Jelalpore, Mymensing
18.	Chittagong	Chittagong	Chittagong, Arrakan, Bulloah (Noakhali), Tippera
19.	Assam	Assam	Assam, N.E. Rangpur, Sheerpur, Sylhet.
20.	Allipore (Alipur)	Alipur	Backergunge, Jessore, Suburbs of Calcutta, 24-Parganas.
21.	Cuttack	Cuttack	Cuttack, Khurda, Balasore, Midnapur, Hijili.
22.	Burdwan	Burdwan	Burdwan, Jungle Mehalas, Hooghly

the Atlas of India. George Everest had emphasized the need of this link-up and therefore, had asked Bedford to keep him informed of his fixed points so that he might be able to intersect them from two or more of his great trigonometrical principal stations, and thus would be able to correct and finally link up the two different surveys and so provide authenticity to his revenue surveys.[121] Similarly, he asked R. Wroughton as well about the list of his fixed points for the purpose when Wroughton was surveying Mathura districts. Everest wanted to supply him with correct latitudes and longitudes of the 'fixed' points together with their azimuths.[122] J. Fordyce in 1838 for his Agra surveys, and others like him, also sought and got help in this direction from George Everest. Fordyce even proposed in his report that the practice of making detailed circuits for revenue maps, which had proved time-consuming, might be discarded in favour of the astronomical fixed points being linked up with the points of trigonometrical survey stations over the whole of the district. This would control and check the distortions of work more effectually.[123]

By 1837, the results of the Allahabad Conference tremendously affected the topographical surveys and mapping. C.R. Markham was of the opinion that the topographical details of the pre-Allahabad Conference Revenue Surveys were tolerably well executed.

To expedite the revenue surveys, the number of existing revenue survey parties was increased from four to nine between 1833 and 1836.

Between 1833 and 1842 the districts mentioned in Table 6 were revenue-surveyed.

In 1842, after the Afghan war, retrenchment became necessary with the result that all those surveys were closed down. Unfortunately during the mutiny of 1857, most of the district maps and records of the North-Western Provinces were destroyed,[124] except the correspondence records preserved at the Board of Revenue, Allahabad.

The surveys that were carried on between 1822 and 1832, the period before the Allahabad Conference, had been very time-consuming and, therefore were slow. These were very expensive also as the then surveyors were after quality and not quantity; the standard of accuracy demanded was very high, and the time spent on surveys of topographical details and mapping, etc., added to the overall expenses.[125] On the other hand, the revenue surveys that were conducted after the Allahabad Conference of 1833 up to 1837 came under the clear-cut rules introduced as a result of the Allahabad Conference

TABLE 6

Agra	-	1837-9
Aligarh	-	1837
Allahabad	-	1838
Azamgarh	-	1835-7
Banda	-	1840-1
Bareilly	-	1833-7
Benares	-	1839-40
Bijnor	-	1833-41
Budaun	-	1822-34
Kanpur	-	1840
Chittagong	-	1835-41
Dehra Dun	-	1838-9
Delhi Territories	-	1824-40
Etawah	-	1838-9
Farrukhabad	-	1833-9
Fatehpur	-	1839
Ghazipur	-	1839-41
Gorakhpur	-	1836-8
Hamirpur	-	1839-40
Jaunpur	-	1839-40
Jalaun	-	1841-2
Mainpuri	-	1837-9
Mathura	-	1835
Meerut	-	1831-6
Moradabad	-	1834-41
Patna	-	1841-3
Sohagpur and Ramgarh	-	1842

in the North-Western Provinces.

Another set of orders, regarding the layout, projection, compilation and engraving of the standard topographical quarter-inch Atlas of India, were issued in October 1823. In brief, the orders stated that the entire work involved in the project was the responsibility of John Walker, the well-known engraver and map-publisher of London, entrusted with the job of bringing out the Atlas sheets by the Court of Directors; but the material for this was to be prepared and supplied by the East India Company in India. For this job, only such surveys were to be considered which the Surveyor-General thought to be worthy of the high standard set for it. The Directors also made it in a way mandatory that the scientific control of the surveys useful for the Altas was to be through their link-up with the great trigonometrical triangulations, being spread out all over India. The triangulation by the Great Trigonometrical Survey which was to be extended ultimately to the whole country was to cover first those areas which had already been surveyed. Thus, priority was to be given to the Bengal Provinces.[126]

To obviate the possibility of any delay in engraving etc., because of the material received from India being incorrect for incorporation into the Atlas, the Directors made the Surveyor-General responsible for it by ordering him to remark upon the merits of each survey being passed on to them.[127] This put the time and ability of the Surveyor-General to a severe strain. When George Everest was on leave to England between 1825 and 1830, the authorities in England had agreed with him that trigonometrical survey should take priority over all other surveys in India. They impressed upon the Surveyor-General that the points upon which the maps of the Bengal Presidency were to be compiled for the Atlas, must have triangulation for their basis and 'expressly prohibited all surveys in which measured distances by perambulator or other means are checked by observations for latitude and longitude'.[128] As trigonometrical survey was the only checking source of the correctness of the surveys, Everest was expected to bring the Great Trigonometrical Surveys to a speedy conclusion so that he would be in a position to expedite the submission of material for the Indian Atlas, all accurately tied to the triangulation.[129]

The Directors had given priority to the completion of triangulation of Bihar and Bengal because they were of the opinion that the geography and topography of these provinces had already been very accurately delineated earlier by James Rennell. Everest, however, very much doubted the mapping value of the old surveys, even when connected up by the triangulation of the Great Trigonometrical Survey. He was of the opinion:

> ... much of the work executed by ... the Revenue Surveys will ... be susceptible of ... combination, and it is certainly to be desired that work of that quality should be made available. There may, perhaps, on the other hand, be materials which will not serve the purposes of the atlas sheets unless more or less voilence be resorted to, to fit them into each other in their proper places. ... All materials which are not radically bad may be turned into account.[130]

Till the great triangulation was completed, no new topographical surveys, after the completion of the Lower Brahmaputra survey in 1834, had been undertaken in the Bengal Presidency.[131] The Surveyor-General was, however, of the opinion that the revenue surveys, conducted before the Allahabad Conference decisions were made applicable, would be more susceptible to adjustments required to make them useful for the Atlas.

About the use of Post-Allahabad Conference revenue maps for the Indian Atlas sheets, C.R. Markham was of the opinion:

> ... The consequences of this (Allahabad Conference) resolution have been disastrous. The surveys of the North-West Provinces were made at a galloping rate each season, owing to the pressure of the revenue officers, who wanted to complete the settlement. The result was that the maps were the merest and most inaccurate, skeletons, while topographical details were altogether omitted. The surveys were confined to the actual definition of village boundaries, and the work on opposite sides

of a river were never even connected. As geographical material they are perfectly useless. These were the materials from which the geographical maps on a scale of four miles to the inch were lithographed at Allahabad, after having been reduced by native draftsmen. They have no trigonometrical points, and no basis of any kind; yet from these maps the sheets of the Indian Atlas have been filled up.[132]

Markham, perhaps, was not accurate in levelling these allegations. It seems he was not aware that sufficient trigonometrical data was available for connecting purposes, after 1844. R. Wroughton, then, had them adjusted and suitably connected to the triangulations of the Great Trigonometrical Survey, re-lithographed at the Government Press at Calcutta and it was only then that the lithographed copies were sent to England for incorporation into the Indian Atlas sheets.[133] Similarly, Delhi maps which were surveyed by T. Oliver, J.H. Simmonds and William Brown during 1824-6, 1830-2 and 1840 respectively, were re-compiled into a single map in 1848.[134] This was also sent to England for use in the Indian Atlas, along with other district maps, on the scale of 4 miles to an inch. These were corrected by the Great Trigonometrical Survey.

District maps of the Lower Provinces were also compiled in a similar manner. Some maps could not be connected to the trigonometrical network as it had not reached the area at the time. In some instances, it took years to cover the region.

In 1831, the Directors had ordered the Surveyor-General that all the district officers should be provided with maps of their districts showing district boundaries. Similar maps for Revenue Collectors and judicial officials should also be got ready as early as possible.[135] Most District officers did not have authentic records or even an idea about the actual limits of their districts. In due course, district maps were compiled and lithographed for the use of officials and general public.

In the meanwhile, the authorities at England were eager to have all district maps as early as possible. This made the Governor-General, rather panicky. To his discomfiture, the Surveyor-General was found almost always away from the headquarters at Calcutta, whenever there was a need for his advice. Actually, the Surveyor-General, George Everest, was away personally guiding the Great Trigonometrical surveys in the Central India. This led to great inconvenience to the Government. The situation became acute when the new Governor-General, Lord Auckland, took charge in 1836. The work in the Survey Department had almost come to a standstill during the long absence of the Surveyor-General from his office at Calcutta. The Governor-General, like his predecessor, showed great interest in the surveys and used to become very much annoyed everytime when, in the absence of the Surveyor-General, no one could explain the policy of the Department to him clearly, or could reply to his queries.

To obviate this difficulty, the Supreme Council formed in 1837, a Survey or Map Committee at the Presidency,[136] primarily to assist the Government in the absence of the Surveyor-General on field duties. The blunder committed in this was that the Committee was constituted of non-professional members, barring three military members who had experience of very few, isolated and scattered surveys; though efforts were made later to involve the Deputy Surveyor-General, who was at that time working as Superintendent, Revenue Surveys, as well, with his headquarters at Allahabad. He was later, posted to Calcutta: and his head office was also shifted to Fort William.

However, before this could happen, in 1837, certain duties were assigned to the Survey Committee by the Supreme Council. The first duty was to compile a catalogue of all maps found at the Presidency offices, including the Surveyor-General's office. *The Register of the Maps to be found in the various Offices of the Bengal Presidency,* was thus completed in 1838. Next year, it was published in two volumes. In 1842, a supplementary volume was also brought out. It included additions of Calcutta records, and lists of materials that were transferred earlier from Bombay and Madras, when the Deputy Surveyor-General's Offices there were closed down by September 1833.

By and by, the activities of the Survey Committee began to clash with the functions of the Surveyor-General. While assigning duties to the Survey Department Surveyor-General was not consulted as he was busy with the work of the Great Arc in the upcountry, nor was James Bedford, the Deputy Surveyor-General, aware of the functions of the Committee, though he was ordered to draft regulations for the control of the map-drawing office. The irony of the situation was that the Council did not recognise or just failed to recognise that the policy, regarding the Great Trigonometrical Survey, being followed by the Surveyor-General, George Everest, had the fullest backing of the Court of Directors, and any interference in it, even by the Supreme Council in India, or deviation from it, without their prior approval, would not be to their liking. The Supreme Council wanted to hasten the compilation of maps from the existing surveys and in their zeal, perhaps ignored the danger of the consequences of interfering in a technical and professional department, apparently on non-professional advice. They passed the following resolution:

The attention of the President in Council has been strongly drawn to the want of system which exists in respect of surveys and the preparation of maps, in the different departments of the public services whereby the expensive operations which have heretofore been undertaken by Government for such purposes have not been made to yield the full benefit which would be derivable from them under a good system....

To obviate this evil, the President in Council resolves that there shall be one office in India, viz., the Surveyor-General's office, in which all original maps and fieldbooks, with the memoirs relating thereto, shall be deposited.... There is ... according to the ... Court's orders ... [vide their] Despatch to the Governor-General in Council in the Military Department dated the 3rd of June 1814.

... in the absence of the Surveyor-General on detached service, the ... office at Calcutta should be placed under the management of an officer competent to carry on ... The Governor-General has suggested that Captain Bedford ... be employed in this situation....The Governor-General has likewise suggested that the ... Survey Committee may be made useful in maturing this scheme of...systematic record of all surveys.... The present drawing establishment ... must in some degree be enlarged.[137]

Soon after, the Council, in their further resolution, established another committee, called the Geographical Committee, with more specific authority.

... The residence of every future Surveyor-General should be fixed at the Presidency,... leaving such work as the Trigonometrical Survey to be prosecuted by a subordinate officer....

The local head of the Survey office should be assisted by a Committee of capable officers with whom he may consult on the collection and application of materials, and on the general improvement and extension of the new system. The Governor-General in Council gladly avails himself of the valuable aid which ... H.T. Prinsep has consented to give as President,...[138]

One of the recommendations of the new Committee was absurd because it helped to squander away the finances. The Committee recommended that all the materials of whatever nature that had been accumulated at Calcutta, including those transferred from Bombay and Madras, relating to those regions of which Indian Atlas sheets had not been compiled till then be compiled on quarter-inch scale. If the material consisted of geographical or revenue maps on smaller scales, enlargements from them be made, and, in case, the surveys or maps were on larger scales, these were to be reduced on quarter-inch scales, for use in the Atlas sheets. Thus, all the revenue maps, including 8-mile revenue maps which were then being lithographed and geographical maps on smaller or larger scales, even on different projections, were to be converted into uniform quarter-inch scales, then lithographed, and copies of them were to be sent to London. They also recommended that the materials for the Atlas Sheets Nos. 119-21, covering areas of Bengal, be first suitably connected in India (at Calcutta) with the sheets of the north-east frontier with the help of the fresh materials being compiled and reduced (or even enlarged) for the purpose, before these were to be despatched to London, even though this job was specifically to be done by experts engaged with the Atlas business at London. The strangest aspect of the whole recommendation was that the Committee were ignorant of the fact that there was no possibility of the Great Trigonometrical Survey triangulation reaching that area to enable them to connect with it their compilation for many years. Thus for all intents and purposes, the authorities showed lack of critical professional knowledge to avoid waste of labour and money involved in making copies of every scrap of material, even when it was of no possible geographical value. But, on the other hand, some of their recommendations were to real worth such as the allotment of some percentage of the work of the Government lithographic press to the Surveyor-General's work.[139]

Thus we see that the schemes enunciated by the Supreme Council in India and the expressed wishes of the Court of Directors, in many instances, were at variance with each other and created more problems instead of solving them.

It seems that these matters were foreseen, by James Bedford in 1840, when he declined the offer of joining as an *ex-officio* member of the Committee. He wrote a long letter to the Government. His lengthy letter was sufficient to expose the inept notions of the Supreme Council, so far as the technical aspects of the mapping material, its accumulation and the knowledge of the state of such material and its utility and availability were concerned. He stated:

Having seen my appointment as member of a ... "Map Committee", respecting which no communication had been made to me by Government, I applied without delay for the documents connected therewith.

It would ill become of me to offer any remarks on a measure which it has pleased His Lordship in Council to adopt, however unexpected on my part, and mortifying to my feelings.... After an attentive perusal, however, of the proceedings of the Committee, objections still stronger than those urged against the immediate compilation of a map of Afghanistan in this office ... appear to apply to the proposed "Compilation Scheme" of the Committee....

Even were the proposed scheme unobjectionable in this respect [like enlargement and reduction, as they might be and also of any map on any projection], and also in point of materials, ... it would still ... involve the necessity of detaining original matter in India, contrary ... to the repeatedly expressed wishes of the ... Directors, who in June 1828 recorded ... that the various unsuccessful attempts to execute an Atlas in India have proved the main obstacle to the timely publication of their own Atlas sheets.

At the same time they conveyed their positive injunctions to the Surveyor-General that the latitudes and longitudes of all surveys sent home shall be attested by him, an injunction which seems directly opposed to the proposed scheme, and for want of the above verification maps ... have been rejected for the Atlas....

In May 1830 the Hon'ble Court directed minor series of Triangles to be based on Mr. [Joseph] Olliver's longitudinal chain from Seronj to Calcutta,... to complete the province of

Bengal. Again in 1833 they remark that sufficient points only are wanting to enable them to finish the Atlas sheets of Bengal and Behar, and in October 1835, repeat their call for triangulation as far as done. It may probably be thought singular that the above points were not brought to the notice of the Committee....

Since I have had ... charge of the Surveyor-General's office, few of the maps ... (those from the Madras surveys excepted) have been accompanied by either fieldbooks, memoirs ... astronomical observations, or other documents, to establish their value....

As a member, therefore, I should be likely ... rather to embarrass them than to aid their proceedings, or become personally responsible for measures in which I might not concur.

During the short time, however, I am likely to continue in office, I shall at all times be ready to give my opinion on any point referred to me—to receive the suggestion of the Committee—and to the best of my power to carry through the ... wishes of Government. But ... I trust I may be permitted ... to decline becoming a member of a body, where my voice (as one of a member) would be in an inverse ratio to the responsibility of my official situation....[140]

James Bedford conveyed his sentiments to the Surveyor-General at the same time, enclosing a copy of the above letter, thus:

... to hurry on the compilation of so expensive a work as an Atlas of India, with materials so imperfectly verified as a large proportion of those in this and other public offices, seemed ... but little likely ... to serve the cause of geography, or to add to the reputation of the office.[141]

Lengthy correspondence ensued with the Supreme Council and the Geographical Committee on the one hand and the Deputy Surveyor-General on the other. Bedford passed on the copies of this correspondence to the Surveyor-General, who wrote to the Government, in clear and forceful language, supporting his Deputy. He further stated that his own earlier suggestions were ignored by the Council. He even went on to say that the Committee was ignorant of the elementary principles of surveying and he had, therefore, in 1832 asked that the professionals should take charge of topographical and geographical work, while the Madras and Bombay map depots be shifted to Calcutta.[142] Finally, the Surveyor-General asked the Council that the whole matter be put before the Court of Directors, which was done. The reply received from them vindicated the stand taken by George Everest and his Deputy Surveyor-General, James Bedford. The interference of the Council into the technical affairs relating to the Atlas was minimised and the Survey Department, henceforth, carried its job in this respect smoothly and expeditiously. The Court had stated:

The institution of a Geographical Committee for ... compiling maps of different portions of India must fail of ... any beneficial result, if it does not occasion injury to the service, and lead to an unnecessary expenditure of the public money.

The Committee is composed of individuals whose time is fully occupied by the duties of their own appointments, and who, besides, have never been engaged in the Survey Department, and cannot therefore ... possess that particular knowledge which is essential to the accurate compilation ... of maps from materials derived from various sources—arranged on different scales—and of which the data are uncertain and occasionally erroneous....

Neither are we surprised that ... Major Bedford entertained strong objections to the plan, and ... declined to be a member of the Committee. He was well aware ... that the plan involved a serious departure from our former instruction, ... [and] that it would seriously interfere with the progress of the Atlas in this country without furnishing any adequate substitute....

It is our earnest desire to carry on and complete the delineation of the Geography of India in such a manner as that it shall be final.... With this view we have resolved to engrave no more sheets of the Atlas unless their contents shall be based upon triangulation.... Combining the different surveys into sheets corresponding with the ... index sheet is to be performed in this country, where the engraving of the whole is also to be executed.

The great desideratum now is continuance of the subordinate series of triangles connected with the meridional arc, which, by fixing the exact positions of a number of places in tracts that have been already surveyed, will allow of great portions of Hindoostan and of the Deccan being finally incorporated into the Atlas....

It is a matter of regret to us that you should have occupied yourselves with ... compiling maps which were not immediately required, ... and ... of portions of the country, the positions of which have not been finally ascertained by triangulation. For ... any map which might be required for political or military purposes, ... every means... should of course be made available, but to... make maps of the neighbourhood of Calcutta and of the Bombay territories appears to have been quite unnecessary....

Any expenditure which may be incurred in the compilation of maps, other than those performed in the Surveyor-General's office and under his immediate responsibility, ... should immediately be discontinued. ... Whatever additional aid may be ... necessary to enable the Surveyor-General to carry on satisfactory ... should be given to the extent of your ability.[143]

The year 1879-80 was another landmark in the evolutionary history of the topographical surveys, when the first rough topographical survey of all India was nearly accomplished as extensive chains of triangulatory measurements had by that time been spread all over the Indian subcontinent; and the principal triangulation was now fast approaching completion, which was reported in the Annual Report of the Survey of India for 1879-80, as detailed later in this Chapter.

As already noted, the orders for the standard quarter-inch Atlas of India, generally called the Indian Atlas, were issued by the Court of Directors in October 1823.

The whole thing was planned to provide a quarter-inch topographical map of the country in 177 sheets, measuring 38 inches to 24.4.[144] This plan was based on the scheme first mooted by Aaron Arrowsmith (1750-1823), who was a friend of James Rennell and A. Dalrymple. Arrowsmith had as early as 1804 published a map of India which was followed by another, a larger one in nine sheets in 1816. By 1822, he brought out a map of south India, on quarter-inch scale, based on William Lambton's triangulation which served as model to the initial sheets of the *Indian Atlas*, which were published from 1827 onwards. But before Arrowsmith could complete all the sheets, he died. The work was then entrusted to J. Horsburgh and then later to John Walker.[145] These sheets for many years were being finally compiled, engraved and printed in England, from the material supplied by the Survey of India, certified by the Surveyor-General to the effect that these were fit to be incorporated into the Atlas. The Directors had specified that the Atlas must be connected with the principal triangles of the Great Trigonometrical Survey. It was in 1864, that the Indian Atlas Sheets were reduced in size, i.e. one quarter of the initial size projected; and many of the old full-sized sheets were revised and re-engraved as quarter sheets so that its publication might closely follow the manuscript compilation. From 1867 onwards, the sheets were engraved and published from Calcutta to avoid inconvenience and unnecessary delay. 'In 1870 the first quarter sheets of Indian Atlas engraved in India (No. 87 S.W. Lucknow, and No. 125 S.E. Sylhet) were issued....'[146]

Before the sheets started being engraved and published regularly in India, the first specimens of the engraved Indian Atlas sheets were submitted by the Surveyor-General for inspection and approval[147] to the Government of India. The Governor-General in Council was highly satisfied with them.[148]

By 1905, on the re-organisation of the Survey of India, all the work on the Atlas of India was stopped; but by that time, the whole of India had been mapped on the quarter-inch scale.

It is a well recognised axiom in all progressive countries that a good general map of the country is essential for purposes of efficient administration, both civil and military. The scale adopted for maps of such a general atlas of the country varied slightly in different countries. In India, as in England, the 1-inch scale had long been agreed upon as the most suitable for the purpose. But still, because of varied reasons, the Atlas of India was visualised on a quarter-inch scale, although this was never recognised universally as the scale suitable for topographical mapping. Upon whatever scale the part of the country might have been surveyed earlier, efforts were made, as and when feasible, to prepare standard sheets on 1-inch scale.

We have seen that the Allahabad Conference of 1833 was a landmark in the evolution of Indian topographical surveys.

In Bengal, which included at that time Bihar and Orissa as well, the post-Allahabad-Conference maps that generally commenced from 1834, were completed by about 1865. These were surveyed by Revenue Survey parties. As the country was almost a flat cultivated ground, the survey was entirely based on traverses, no triangulation being executed, but stations of the Great Trigonometrical Survey were being connected with these traverses wherever possible. The method of survey adopted was to make village maps, at first, village by village, and subsequently, compiled in groups, on the 4-inches scale. But there were following exceptions to this general rule:

1. The districts of Hazaribagh, Singhbhum, Ranchi and Garhjat state districts of Chota Nagpur, where topographical surveys, on the 1-inch scale, were made during 1856-72.
2. In the Tributary Mahals of Orissa, a topographical survey was made between 1850 and 1863, on the scale of 1 inch=2 miles.
3. In the Chittagong Hill Tracts and in Hill Tippera, a rough reconnaissance, on the 1-inch scale was carried out by a Revenue Survey party.[149]

The revenue surveys in Bengal of the period in question mentioned above were not of much use when topographical details were required, because these were executed in the early days of topographical maps surveyed by Indians who were trained to survey village boundaries only.

A decision was taken to commence Cadastral Surveys in India in place of earlier Revenue Surveys in 1871, as it was considered the cheapest and best method in the long run. The surveys were conducted on a 16-inch scale. Though begun in 1871-2, these could be introduced in Bengal only in about 1886.

So far as the state of topographical mapping of the North-Western Provinces and Oudh was concerned, the original surveys in the plains of the post-Allahabad-Conference Period consisted as in Bengal, of village surveys on the scale of 4-inches to a mile, conducted by the Revenue Survey Department. These surveys date back to 1833, and were practically completed by 1870. It was from these maps that topographical maps were compiled.

As already mentioned, maps and records of the original surveys of the sixteen districts listed in Table 7 were destroyed during the uprising of 1857.[150]

These districts were later ordered by the Government of India to be resurveyed on the scale of 2 inches to a mile, with village boundaries as well being shown

TABLE 7

1.	Muzaffarnagar	9.	Etawah
2.	Meerut	10.	Kanpur
3.	Bulandshahr	11.	Jalaun
4.	Aligarh	12.	Fatehpur
5.	Etah	13.	Allahabad
6.	Farrukhabad	14.	Azamgarh
7.	Mainpuri	15.	Basti
8.	Shahjahanpur	16.	Gorakhpur

thereon.[151]

The maps of the remaining districts were prepared on a 1-inch scale. Because of the time-lag, these had become out-of-date by the time these were compiled.

It was in about 1872 that Cadastral surveys were introduced in North-Western Provinces for settlement purposes. These surveys were based on scientific system of traverse on a 16-inch scale. The trijunctions of village boundaries were properly fixed and preserved. Topographical maps were compiled from them in the form of standard sheets, on a 1-inch scale and of sizes of 30 minutes of longitude by 15 minutes of latitude each, which were later in 1905 reduced to 15' of longitude X 15' of latitude, just half of the old standard sheets.[152]

As the Cadastral surveys were introduced in 1872, not much progress was made in that direction by 1875-6.

In the hill-region of the North-Western Provinces, Tehri Garhwal, a reconnaissance survey had earlier been carried out in 1851-3 on a small scale of 1 inch to 4 miles.[153]

So far as Punjab of those days was concerned, the hilly region in the North-Western Himalayas, has mostly gone to Pakistan. In Jhelum and Rawalpindi, the topographical survey was made during 1848 and 1865 while in Hazara district, the topographical survey was carried out by a Revenue Survey party and, therefore, topographical features were not properly brought out.

In the plains of Punjab, the Revenue Surveys, were executed between 1846 and 1880 on the scale of 4 inches to a mile. These maps were later issued as district maps.[154] In August 1874 orders were issued for the introduction[155] of Cadastral Surveys in the Punjab and these began with the Rohtak district.[156]

To complete the survey of the north Indian scenario, we may now examine the mapping activity in the north-eastern region.

The earlier maps in Assam, as in the case of other provinces of the Bengal Presidency, were based on surveys undertaken by the Revenue Surveys parties; surveyed, as usual, on the scale of 4 inches to a mile. These were applicable to the cultivated portions only of all the districts between the years 1853 and 1893. The results, as in the case of other districts, were published in the form of standard sheets on 1-inch scale.[157]

The survey of the hilly portions of the province was undertaken by the topographical survey parties from time to time. First, such topographical surveys were carried out on the scale of 1-inch to a mile in the Garo and Khasi in 1864, and on 2-mile scale in the Jaintia hills, during 1879. The tracts, adjacent to the borders of the regions inhabited by various tribes, were usually in a disturbed state. Ostensibly to settle their border problems, military expeditions were sent there by the English from time to time. These gave opportunities to reconnoiter and survey the region. Such surveys were on scale of 4 miles to an inch.

The opportunity to revise such surveys came, as in most other provinces, when Cadastral surveys were introduced. Cultivated portions of the districts of Sibsagar, Lakhimpur, Nowgong, Darrang, Kamrup, Goalpara, Sylhet and Cachar of Assam were cadastrially surveyed during the years 1853 to 1875.

Much of the region was surveyed by parties attached with the Assam Valley Series, the Cachar Secondary Series, the Eastern Frontier Series, etc., of the Great Trigonometrical Surveys.[158]

In the Central Provinces, the work of the Great Trigonometrical Survey was completed in 1876 having been covered by four series of triangles through Jabalpur and Sambalpur. As the topographical survey in the region succeeded[159] the trigonometrical surveys, the earlier one could thus be linked with the latter. The topographical survey was begun in 1862 and in 1873 the operations in the Central Provinces were concluded.[160] Beginning with 1862, till 1877, both the revenue and topographical parties were employed and each surveyed according to its own practices and training. The question of the mapping of the Central Provinces presented the greatest difficulties. The nature of the topography differed considerably from other provinces of the Bengal Presidency. Here, instead of the unbroken plain covered with cultivation, the land presented an intricate mixture of jungle-clad hills and cultivated plains. The maps resulting from these surveys were published on the scale of 1-inch to a mile, whereas revenue surveys, which began in 1863, substituted the topographical surveys for open and cultivated areas. The parties carefully traversed the village boundaries with theodolite and chain and filled in the details of the flat ground, on the scale of 4 inches to a mile, by plane-tabling.

Turning to Central India and Rajputana Agencies, the topographical surveys on a 1-inch scale of the native states were started in 1860 and were carried on until 1886, with the exception of a small portion of the desert part of Rajputana which was completed by 1891. In the Central India and Rajputana Agencies, as per general policy, in the initial stages, only topographical survey parties were employed to survey on the usual 1-inch scale. But much later, in *Khalsa* or revenue-paying areas of Rajputana, Cadastral surveys were introduced by 1860. In all these states' surveys, village boundaries were not required.

In the Central India, Gwalior and Central India survey party commenced work in an area north of the 24° latitude, bounded—'... on the west by the Rajputana Agency, and on the east by the districts of Jalaun, Jhansi and Saugor and comprised Datia, and parts of Gwalior, Indore, Jaora, Khilchipur, Rajgarh and Tonk States.[161] In 1862, a party was organised to survey Rewah and Bundelkhand which was completed in 1871. The party was then transferred to Bhopal and Malwa and surveyed the country lying south of 24°N latitude and north of 22°30'N latitude and the Narbada river. By 1882 areas—'... comprising the states of Bhopal Narsinghgarh, Dewas, Jaora, Ratlam with portions of Gwalior, Indore, Jhabua, Khilchipur, and Rajgarh were surveyed on 1-inch scale.'[162] During 1871-2, a Khandesh and Bombay Native States survey party surveyed part of the Agency—'...lying south of 22°30' [N. latitude] bounded on the east by Nimar district, and on the south and west by Khandesh and the Rewa Kantha Agency, embracing Barwani and Jobat, with portions of Dhar, Indore, Gwalior, Dewas and Ali Rajpur.'[163] And finally, in 1870-1, the Rajputana Topographical Survey party surveyed part of Gwalior and Indore.

In the Central India States Agencies, revenue surveys in the western portion of Bundelkhand comprising the states of Orchha, Charkhari, and Samthar, and the smaller states falling within Hamirpur district were surveyed on a 1-inch scale in 1855-7 and 1860-2. Parts of Jalaun and Jhansi districts were surveyed on 4-inch scale by a revenue survey party during 1852-7 and 1859-60.[164]

In the Rajputana States Agency the Great Trigonometrical Survey, during the period under review, was extended only to parts of Rajputana. The topographical survey of the entire country was carried out by the Survey of India between 1855-91. The Cadastral survey, which began in about 1860, was confined to the *Khalsa* or revenue paying areas only.[165]

In Bombay, the revenue or cadastral surveys, which were the basis of land assessment, date from the reign of Akbar, the Mughal Emperor. These were based on the revenue system introduced by Todar Mal. At the time of Shahjahan, this system was extended to the Deccan. The first survey of which records exist, were those undertaken by the rulers of Bilaspur at the end of the sixteenth century. As the territory came under the sway of the Marathas, these measurements were partly revised. During the British administration, it was in 1835[166] that a systematic land survey was introduced[167] by George Wingate. This system was considered '...the best [for revenue purposes] in India. But the importance of making the Survey available for general use in the compilation of maps was lost sight of, and the topographical surveyors in Gujarat have found great difficulty in making use of the revenue village maps.'[168] The Local Government, in the Bombay Presidency, had its own Revenue Survey Department, and, therefore, the revenue survey parties of the Survey of India, did not work there.

About the Bombay Presidency, the Surveyor-General, in his detailed letter of 1871, stated:

> ... We have there neither proper professional surveys, nor published Atlas sheets, nor any other sort of reliable map of a large portion of the Northern Division or districts. There is consequently but one remedy ... and that is, to take up the Non-Regulation or Native States by the usual topographic system of survey, on the 1 scale,... [in August 1869] ... and like-wise by a Revenue Survey party or two, on the Bengal model, capable of combining Revenue with Topographic survey delineation, for any Regulation or revenue paying Districts, not at present occupied by the Bombay Revenue Settlement Commissioner's Establishments.
>
> ... until the Imperial system of survey ... is extended to Bombay, no progress whatever will be made towards the Geographical development of that important part of India....
>
> ... But without proper and adequate surveys, there is in fact no royal road to the Geography or correct Topographical delineation and representation of the Bombay Presidency, than there is to any other part of India, or other countries....[169]

In Bombay Presidency, the surveys began first with those of Cutch, Kathiawar and Gujarat between 1868[170] and 1870. In November 1882, the Surveyor-General of India stated that from the very commencement of Gujarat topographical surveys, utmost possible use was made of old revenue surveys.[171] These were concluded by about 1884. Although the surveys were conducted on a 2-inch scale, the maps of Native States were published on a 1-inch scale, and those of British territory, were published on the scale of their survey, i.e. 2-inch scale. In the maps of the British territory, village boundaries were inserted by taking them from the maps of the local revenue survey.[172] In the Khandesh area and the Deccan, the survey began in 1871. There was much controversy about the value of these surveys leading, in 1875, to a Survey Conference at Poona, which did not come to any very satisfactory conclusions.[173] By 1896,[174] the entire Bombay Province was mapped. Only the forest covered part of Kanara, was surveyed on 4-inch scale.

Revenue surveys in different parts of India in the nineteenth century were unfortunately conducted on different principles. The models furnished by the Bombay and Madras Presidencies were both remarkably good. 'The ideal survey, while furnishing complete information for settlement purposes, should be executed throughout on accurate principles, and supply materials for compiling maps for general use.'[175] The Bombay revenue survey was admirable and was considered the best for fiscal purposes; but it did not supply enough material for the preparation of maps for general use. The Madras revenue survey was free from this drawback; it complied with all the demands made on a revenue survey in those days.

By the end of 1855, when the proposal was sent to the Government of India for approval, no regular general survey had ever been made in Madras Presidency.

> ... the Board of Revenue at Madras were first disinclined or at least indifferent to the geographical aspect of the question, and objected to the survey being conducted in connection with Colonel Lambton's triangulation. They urged that they only required field maps for fiscal purposes; and that geodesical operations would cause extra expense and delays, which they strongly deprecated. Major Thuillier, on the other hand, in a letter dated May 6th, 1857, represented that no general survey of the Madras Presidency ought to be commenced without full and ample precautions being taken for making the materials subservient to the general purposes of geography. At the same time he showed that the utmost facilities existed for connecting the Revenue Survey operations with the Trigonometrical points of Colonel Lambton.[176]

The Court of Directors had earlier approved the proposals put forward regarding the topographical survey and the utilization of early revenue maps for the purpose by Colonel H.F.L. Thuillier, the Surveyor-General of India, in their letter dated 17 December 1867. In connection with his proposals for the utilisation of the available geographical information, for the preparation of maps, both general and topographical, of the Madras and Bombay Presidencies and for topographical surveys for such of the districts as still remained, to complete the Atlas of India, Colonel Thuillier reported:

> ... we have both new surveys of a proper character going on, and also the whole of the Atlas sheets of the Southern Peninsula from the old Military Institution Survey, so that there is no object whatever in Madras of compiling and putting together for temporary purposes, desultory and incongruous old materials, which would be tedious as well as expensive and of no use....[177]

Captain F.J. Priestley was appointed the Superintendent of the Madras Revenue Survey[178] on 18 August 1857. The revenue surveys that were approved by the Court could be described thus:

> ... it is designed to show all the principal variations in the surface of the soil, such as hills, jungles, woods, channels, tanks, topes, houses, cultivated and cultivable land, whether *nanjai* (irrigated) or *panjai* (unirrigated), and the area of each field. The village maps are on a scale of 16 inches to the mile, the taluk maps on 1-inch to the mile, and it was intended that, from these materials, district maps should have been compiled on a scale of half an inch to the mile. The work is connected with the Great Trigonometrical Survey....[179]

In the Madras Presidency the revenue or Cadastral surveys of Tinnevelli, Trichinopoly, Salem, Chinglepat, Nellor, Kurnul, Kistna, and Godavari districts were completed, along with special surveys of Nilgiri Hills, Sherveroy Hills, Wynaad, Nellambur Valley Teak plantation estates and of Madras town were taken up in 1856[180] and completed during 1873-4. Surveys of the following districts of Madura, Coimbatore, North Arcot, Cudapah, Ballari, and Ganjam could be completed only by 1893.

The Madras Government had always had a Survey Department of its own, apart from the Survey of India's contribution to Madras surveys. The Madras Survey Department was mainly interested in revenue surveys. We know that in the first four decades of the nineteenth century, the Madras Government had prepared topographical maps of all the districts under its control. Most of these surveys were made by the Quarter Master General's department and the results were incorporated in 23 sheets of the Atlas of India which covered the entire Madras Presidency.[181]

With the re-organisation in 1855-6 and the appointment of Capt. Priestley as the Superintendent of the Madras Revenue Survey, 1857, the Department carried out Cadastral surveys as well since 1856, and since 1858, all the Government villages, except those in the hilly areas, were thus surveyed on a 16-inch scale. These maps depicted the boundaries of every field and all important topographical details. The field-books and measurement-registers were also meticulously maintained.

Between 1867 and 1877, a topographical survey of hilly and forest-clad portions of *zamindaris* of Ganjam and Vishakhapatnam, and Tinnevelli, Malabar, South Canara, Madura, Trichinopoly was conducted on a scale of 1-inch to a mile, though maps of more inaccessible regions were made on a 2-mile scale.[182]

In Mysore, the revenue surveys were commenced as early as 1863 and the settlement was completed by the end of 1901.[183] In the meanwhile in 1874 the Government of India ordered the immediate professional topographical survey of the state.[184] Two survey parties, one under George Strahan, and the other under J.R. McCullagh began surveying on a 1-inch scale. Nandidrug and Ashtagram divisions were taken up by Strahan, and Nagar division was surveyed by the latter. This topographical survey was completed in 1886.[185]

Hyderabad state was included in the Great Trigonometrical Survey. Subsequent topographical surveys were based on, or were extended from, the main series of triangulation.[186]

The survey of the districts of the Nizam's dominions and the Hyderabad Assigned Districts was considered the oldest district survey existing in India.[187] The surveys were commenced in 1816 and continued till 1847. These were mostly on a 1-inch scale. The maps were published as *circar* (an administrative division, like a district) maps.[188] Regular districts were formed in Hyderabad in 1865-6.[189] In about 1876, it was decided to have regular and accurate survey operations. It was first begun in

Aurangabad and then in Gulbarga Divisions. These were settled by the end of 1894.

Berar was surveyed between 1843 and 1850 and between 1856 and 1866, a small portion of it was re-surveyed.[190]

During Henry Edward Landor Thuillier's period (1861-77) as the Surveyor-General of India, and the Superintendent of Topographical Survey, our knowledge of the physical geography of India increased considerably. His studies, surveys and travels were, mostly over the

> ... mountainous and forest-clad regions, or over sandy deserts, frequently in parts never before visited by Europeans. Colonel Thuillier truly says that for frontier expeditions and wild tract of country, the Indian Topographical Survey method of surveying by theodolite and plane table based on the great and minor triangulations cannot be excelled for general accuracy, rapidity and cheapness, if the agency employed is well-trained and trust-worthy....[191]

We have noted the official surveying and mapping activities till 1879-80. There were many publishers outside India who were very much interested in the Indian survey activities and in the publication of maps of India. Some received materials for the purpose through official channels and some had to depend on their own resources.

John Walker succeeded J. Horsburgh, the Hydrographer to the East India Company in 1836 and remained on the job till 1861. It was during his time that Marine Surveys and Charts were transferred to the Admiralty. It was John Walker who edited, engraved and published them. An 1878 catalogue of the maps, etc., of the Indian Surveys available at the India Office, London, mentions 201 such charts,[192] and out of this lot 49 charts pertain to India.

The only private publisher of any consequence, of the period was William Allen. In 1826, in a joint venture as publisher under the name—Kingsbury, Parbury and Allen he published a map of the Western Provinces of Hindoostan, inscribed to Major-General. Sir John Malcolm. During 1846, Allen published the map—*Map of the Western Provinces of Hindoostan, the Punjab, Rajpootana, Sinde, Kabool etc*. Although the Great Trigonometrical Survey had covered the area much earlier, the results were not available to public till then.[193] The complete set of Indian Atlas sheets was also not ready till then for publication.

Thus, though the whole country had been topographically surveyed by 1880 the latest maps were not available to public.

We have also noted how the surveyors attached to various armies, political missions, etc., were making secret traverses of routes travelled by them. Similarly, the British authorities in India engaged or employed the natives for such secret traverses as well, after giving these explorers the minimum required training in the elementary use of basic instruments like compass, observation of boiling points of water with the help of thermometer, chronometers or watches; taking of astronomical observations, after recognising various planets and stars; noting of azimuth amplitude and above all, secretly measuring of distances while travelling over various terrains, with suitable methods, etc. After training, they were smuggled into the trans-Himalayan regions, all along the frontier from the west to north and north-east. Thus, they collected the geographical information of all the neighbouring countries.

The explorers, relevant to the period under examination are listed in chronological order along with the areas they traversed in Table 8.

From the above, we see that most of the sub-Himalayan region from west to east, and also Nepal, Tibet and even portion of Central Asia, were traversed and geographical information obtained, and incorporated into the maps under preparation, both by private publishers and the Government.

While submitting the General Report of the Operations of the Survey of India and of the Surveyor-General's Department, 1875-6, the Surveyor-General stated that

> for frontier expeditions and explorations, where rapid progress is essential; for the Independent Native States in which the use of the chain by our surveyors is ill adapted or objected to; for forest clad, rugged and mountainous country; and for poorly populated, non-revenue yielding and insalubrious or inhospitable districts and States; and, in fact, wherever a first and speedy survey, at a moderate cost, is necessary for administrative, Military and Geographical purposes, the Indian Topographical Survey method of surveying by theodolite and plane-tabling, based on the Great and Minor triangulation, cannot be excelled for general accuracy, rapidity and cheapness, if the agency employed is well trained and trustworthy.

The areas thus surveyed were nearly in

> ... every Province and Native State above the parallel of 16°N. latitude,... but chiefly Native States and non-revenue paying British districts, such as are mountainous or rugged, forest clad, sandy desert, and in many instances, ground [region] fatal to health; tracts which have never been visited or explored by Europeans.[194]

From 1875-6, onwards, till the introduction of new survey methods in 1905, along with the re-organisation of the Survey of India, the Department, while as active as ever in the direction of fieldwork, appeared to have weakened as regards cartography or map-making was concerned.[195]

We know that the Great Trigonometrical surveys were being carried on uninterrupted even during the days when there was retrenchment in every other department. We also know that the topographical and revenue surveys furnished the details required for filling up the

TABLE 8

Sl. No.	Year of Exploration	Name of the Explorer	Route Explored/Surveyed
1.	1865-6	Nain Singh	Kathmandu-Kirong-Tadam-Janglache-Shigatze-Lhasa-Tadam-Mansarowar Lakes-Kumaun.
2.	1866	Mani Singh	Milam (in Kumaun) -Gartok-Milam.
3.	1867	Nain Singh	Badrinath (in Kumaun)-Totling Math-Gartok-Thok Jalung-Indus River-Gartok-Totling Math-Kumaun.
4.	1867	Mani Singh	Totling Math-Shipki-Nilang.
5.	1867	Kalian Singh	Giachuruf-Jiachan
6.	1867	Kalian Singh	Lajun Chunik-Dewchok-Dunkhar-Totling Math.
7.	1867	Kalian Singh	Totling Math-Shipki
8.	1868	Hari Ram	Pithoragarh-Kathmandu-Tulsipur-Baramdeo
9.	1868	Kalian Singh	Dunkhar-Demchok-Rudok-Thok Jalung-Mansarowar Lakes.
10.	1868-9	Mirza Shuja	Kabul-Faizabad-Panjab-Little Pamir-Yangi Hissar-Kashghar-Kurram Pass-Leh.
11.	1869	Krishan Singh	Kathai Ghat-Karnali River-Khojar Nath-Rakas Tal-Milam.
12.	1870	Hyder Shah	Peshwar-Dir-Chitral-Faizabad-Zebak-Dora Pass-Peshawar.
13.	1871-2	Hari Ram	Darjeeling-Shigatze-Dingri Maidan-Kathmandu-Darjeeling.
14.	1872	Hyder Shah	Kabul-Bokhara.
15.	1872	Krishen Singh	Shigatze-Tengri Nur Lake-Lhasa.
16.	1873	Hari Ram	Pithoragarh-Jumla-Muktinath-Tadam-Muktinath-Tribeni.
17.	1873	Atta Muhammad	Jalalabad-Pashat-Dir-Chitral-Mastuj-Baroghil Pass-Tuskhurghan-Yarkand.
18.	1873-4	Hyder Shah	Kolab-Sagri Dasht-Kila Kumkila Yazghulam-Faizabad-Rastak-Khanabad-Baghlan Haibak-Bamian-Kabul
19.	1873	Kalian Singh	Shabidula-Kiziljilga.
20.	1873-4	Krishen Singh	Tashkurghan-Yarkand-kargalik-Guma Khotan-Kiria-Polu-Pangong Lake-Tankse.
21.	1874	Nain Singh	Leh-Tankse-Noh-Tangri Nur-Lhasa-Chetang-Tawang-Udalguri.
22.	1875-8	Lala	Darjeeling-Sikkim-Shigatze-Chitang-Tawang-Shigatze Chumbi-Darjeeling.

outlines formed by the fixed points of the triangulation, the former were generally confined, to the Native States, and non-regulation British districts, non-revenue producing regions. Actually, for many years, the publication of the results of the topographical surveys, other than the Atlas sheets, followed no uniform or systematic plan. Detached surveys were executed and their maps, generally on a one-inch scale, formed separate series, the sheets conforming, not to the limits of a projected graticule, but to those of areas of administration. In other cases, the units adopted were districts or *circars* or groups of districts; and sometimes in the case of maps compiled from revenue surveys, the main traverse circuits from which the traverse framework on the survey was built up.

The revenue surveys were carried out in the more productive and valuable tracts; their scales being varied to suit local requirements and conditions.

Generally speaking, the standard scale of the topographical survey was 1-inch to a mile, but where the work had to be executed very quickly, the scale was reduced to half or to quarter of an inch to a mile; but in British territory, it used to be increased to two inches, and in case of forest surveys, to four inches. The revenue surveys were generally on scales of 4, 16, and even 32 inches to a mile.

We have also seen that, by the end of 1875, these two classes of surveys, revenue and topographical, had covered approximately the whole of the country, including majority of Indian States, except the western half of Rajputana, portions of North-Western Provinces, the Konkans and most of the Madras Presidency where cadastral surveys began in 1856.[196] All this is related to the period after the introduction of cadastral surveys and the consequent preparation of topographical maps of the regions where cadastral surveys had come into operation.

From here, we now move on to the take-off stage of the modern maps, i.e. the end of 1904 when by a resolution, a committee was constituted for making a complete examination of the existing methods and working of the Survey of India Department in all its branches, with special reference to the preparation or revision and the reproduction of the topographical maps of the country.[197] The deliberations of the Committee resulted in the ushering in of the modern maps.

Between the last phase of the topographical maps examined so far, and the formation of the Committee, the progress of the mapping activity could be summed up as follows.

We have seen that cadastral surveys were introduced in Bengal in 1886. All these surveys were the basis of the later 1-inch topographical maps.

In a flat country like Bengal where natural topographical detail is, with the exception of rivers and streams, comparatively scarce and where the military importance of the country is small, such reduced [topographical maps were compiled from reductions of Cadastral maps] cadastral surveys make very fair topographical maps, but in other parts of India where those conditions do not prevail, it has been well proved that topographical maps so compiled must be examined in the field before publication and the topography added to and corrected where necessary, the chief items for which additional work is required being generally found to be hilly ground, ravines or broken ground, banks of streams and the true shapes of village sites.[198]

Until the end of September 1904, the following districts were thus surveyed on a 16-inch scale: Chittagong, Santal Parganas, Singbhum, Noakhali, Muzaffarpur, Champaran, Saran, Gaya, Darbhanga, Monghyr, Cuttack, Balasore, Puri, Tippera, Palamau, Bhagalpur, Darjeeling, Purnea, Bakarganj, Ranchi, etc.[199]

In the North-Western Provinces, where the Cadastral surveys were introduced in 1872, in the post-1875 period the following topographical maps were prepared: topographical maps of Kumaun and Garhwal were prepared in 1886-88 on a 1-inch scale; a map of Dehra Dun districts and the Siwalik Hills was made in 1873-6, on a 4-inch scale. The hilly portions of Mirzapur district were surveyed on a 2-inch scale during 1882-6 on topographical basis; between 1877-84. Saharanpur, Muzaffarnagar, Meerut, Bulandshahr and Aligarh were topographically surveyed on a 1-inch scale; on these village boundaries were also shown.

The following districts were thus surveyed until 1904:[200]

1.	Agra	1876	14.	Garhwal	1877
2.	Aligarh	1883	15.	Ghazipur	1881
3.	Bahraich	1896	16.	Gonda	1900
4.	Ballia	1900	17.	Gorakhpur	1888
5.	Banda	1893	18.	Hamirpur	1876
6.	Bareilly	1897	19.	Jaunpur	1840
7.	Basti	1887	20.	Jhansi	1898
8.	Benares	1884	21.	Kheri	1898
9.	Bijnor	1914	22.	Meerut	1897
10.	Dehra Dun	1884	23.	Mirzapur	1899
11.	Etah	1884	24.	Moradabad	1905
12.	Etawah	1901	25.	Pilibhit	1898
13.	Farrukhabad	1901	26.	Shahjahanpur	1900

As long as these surveys were based on scientific system of traverse, trijunctions of villages were properly fixed. Only such maps were used for compiling topographical Standard sheets on a 1-inch scale.

. The following districts however did not have complete old traverse data which was lost because of the Mutiny of 1857, or other reasons, so these districts for the time remained without topographical maps based on revenue traverses: Budaun, Fatehpur, Cawnpore and Allahabad.[201]

In Punjab, post-1875 surveys were of two distinct categories, one on 1-inch scale in the hilly region, but in some cases reduced to a 2-inch scale; and two because of the need of the revision of the revenue settlement on a larger scale of 4-inches and 16-inches mostly in the plains, from which the standard 1-inch sheets were prepared. The first category of surveys were carried out by the Survey of India parties, based on triangulation and the other first by Survey officials and later, by the local *patwaris*. The topographical maps, compiled from these sheets, were however not reliable and complete.

In the first category were the maps of Kangra, Kullu, Hoshiarpur (hilly portion only) and the Shimla Hill states; and, in the plains, the revenue surveys on a 4-inch scale were published, earlier on a 2-inch scale and then later on, on a 1-inch scale and some of them as standard sheets. From 1884 to 1889 these large-scale village maps were surveyed on what was then called 'square' system, peculiar to Punjab and were entirely surveyed by *patwaris*.

By the close of the period 1875 to 1900, now under study, the following revenue surveys were made in the Punjab:

1.	Ambala	1875-7	5.	Ludhiana	1875-89
2.	Hissar	1875-8	6.	Patiala	1875-89
3.	Kapurthala	1885-6	7.	Rohtak	1876-7
4.	Karnal	1875-8	8.	Sirsa	1875-8

Many of the surveys were a continuation from the earlier period.[202]

We may now have a look at the situation in Assam and Bengal. As had already been stated earlier, Garo, Khasi, Jaintia and Naga Hills were surveyed on topographical basis, on a 1-inch scale, during 1864-79.

The cadastral and settlement surveys of the districts of Sibsagar, Lakhimpur, Nowgong, Darrang, Kamrup, Goalpara, Sylhet and Cachar were not found of much use for preparing topographical maps.

Coming back to the Central Provinces and its maps, we have already noted that topographical surveys of the hilly and forest tracts were commenced in the Central Provinces in 1862 and were completed in 1873. Re-survey of these maps was taken up in 1902-3, just to check the accuracy of the then existing maps.[203]

These existing topographical maps, by the end of the nineteenth century, embraced almost every variety of surveys. Many of them were from old 'Congregated Revenue Survey maps', and 'Main Circuit' maps which were executed about 40 to 48 years ago. These were prepared by revenue surveyors who had good knowledge of the survey of flat cultivated lands but did not have sufficient knowledge of hill surveying. So the topography was represented in them conventionally, with no attempt at accuracy.[204]

For cultivated areas, traverse survey with simultaneous cadastral surveys were carried out, but on different scales, independent of each other from 1884-5 onwards. The topographical maps prepared out of them were not of much use.

The topographical maps of Cutch, Kathiawar and Gujarat were prepared from the topographical surveys conducted during 1868-84.

The Khandesh and Deccan survey parties were at work from 1871 till the end of 1896. Both the surveys were on a 2-inch scale. These were later published on a 1-inch scale.

Similarly, topographical surveys of the hilly and forest-clad areas in Bombay were also carried out. In Mysore, the topographical work was commenced in 1875 and was completed in 1886;[205] and likewise, in Hyderabad, the *'circar'* maps were redrawn, and republished, with additions, during 1884 and 1887.[206]

Thus we have seen the progress of the topographical mapping in the country till the time mapping in the country started being seriously discussed by the Survey Department. The functions and duties of the three departments, the Great Trigonometrical Surveys, Topographical Surveys, and the Revenue Surveys were seriously examined. In 1878 the Surveyor-General wrote to the Government:

> ... the duties of the three Departments [Great Trigonometrical Surveys, Topographical Surveys, and Revenue Surveys] have become gradually intermixed,... There are still certain duties in connection with the geodetic operations of the Great Trigonometrical Surveys, for which special training and study are required. But by far the greater portion of the survey work now in progress and remaining to be done, it is a matter of indifference to which of the three branches of the Department, as originally constituted, the officer undertaking it may belong.
>
> ... The time has thus obviously arrived when it is desirable that the amalgamation of work should be recognised by the abolition of the distinctive appellations by which the three branches of the survey have hitherto been known. As the Department is directly under the Government of India, and its members are liable to serve in any part of India,... it may be simply designated as—The Survey of India;....[207]

The proposal was approved by the Governor-General in Council.[208]

During this very year, another controversy arose which engulfed the Survey of India in serious discussions which had a bearing on its future programmes.

The Board of Revenue, Lower Provinces of Bengal, expressed their opinion that the surveys, necessary for settlement purposes, might be done at far less cost by non-professional agencies. In reply the Surveyor-General of India, General J.T. Walker, wrote a lengthy letter to them when they, as a result of this controversy, discontinued the Dearah survey by professional agency of the Pudda and Megna rivers in 1878. Since the question of total cost was involved, the Surveyor-General examined the problem in detail. According to him

> ... The work of the professional Survey Department under the Government of India may be broadly divided into three classes: 1st, the general survey on the standard scale of 1-inch to the mile, of the hilly, more rugged, and poorer portions of the British districts, and of the whole of the native states; 2ndly, the mouzawar or village survey, on the scale of 4-inches to the mile, of the richer portions of the British districts, as an aid to the operations of the settlement surveys; 3rdly, the cadastral or field survey, on the scale of 16-inches to the mile, of British villages, in supersession of the field measurements of the settlement surveys. Other classes of survey,... do not furnish as good a criterion of the cost of the work ... they need not be adduced at present....[209]

He then proceeded to indicate the nature and cost of each of the three classes of operations, taking his figures and facts from the previous published annual reports for the years 1867-8 to 1876-7. The three classes of operations were as follows:

> 1. *The Topographical Surveys on the 1-inch scale.*
> The survey is constructed by the method of plane-tabling, on a trigonometrical basis in such a manner as to exhibit the positions of towns and villages, the courses of rivers, the general features of the ground and as much detail as the scale will admit of, with a sufficient degree of accuracy for the requirements of a general topographical map on the 1-inch scale. The surveyor goes over the ground with his plane-table, which he sets up at a sufficient number of points to enable him to 'cut in' all the most prominent objects, and to sketch the general features of the country; the number of these points, called 'plane-table stations', will vary with the nature of the ground, being few where it is flat and open, and numerous where it is hilly and rugged. In all cases the plane-tabler actually goes over a comparatively small portion of the ground, the greater portion being sketched in from his stations. He surveys no village boundaries, but only the principal boundaries, viz., those of the British districts and the native states, and their sub-divisions when they have been demarcated beforehand.[210]
> Taking into account the figures from the previous ten years' reports, as stated earlier, the total cost of survey came to Rs. 21-12 (old Indian currency) per square mile.[211]
>
> 2. *The Village or Muzawar Survey on the 4-inch scale.*
> This survey is also constructed by the method of plane-tabling, but on a basis formed by carrying traverses (with theodolites and chains) round the boundaries of the villages, instead of on a triangulated basis. The usual details of the ground are laid down by the plane-tabler, who is also required to delineate the limits of the lands which happen to be respectively cultivated, fallow, or waste at the time of survey. For this reason and because of the larger scales of survey, the plane-tabler has to go over the ground very much more closely than in the Topographical Survey, and his rate of progress is proportionally slower.[212]
>
> The total cost of this class of survey was Rs. 53 per square mile.[213] And finally
>
> 3. *The Cadastral Survey, on the 16-inch scale.*
> This survey is also constructed on the basis of the traversed

boundaries of the villages; the whole of the interior details of the fields are obtained not by plane-tabling, as in the previous instances, but by systematic chaining, which is duly recorded in field-books, and is thus available for future reference whenever it may be wanted.[214]

The total cost, in this instance was Rs.165 per square mile. In this case, the figures for only six years could be taken into account in 1878, as the cadastral survey was comparatively new to Bengal, being introduced in 1871-2 only.

As for the relative intrinsic value of the three classes of survey, a comparative study would show that for a cheap and fairly accurate first survey of India, the first-named survey was considered the best, as it showed the general features of the country and all the most prominent objects on the ground, with a fair degree of accuracy, with errors rarely exceeding 1/20th of a mile. The topographical survey on a 1-inch scale was, therefore, admirably constituted. As a basis for the engraved Atlas of India, on a $\frac{1}{4}$-inch scale, nothing could be more admirably suited. The *mauzawar* or village survey afforded a careful record of the village boundaries and also a valuable check on the field measurements by the *amins*. But the cadastral survey, though it was expensive about three times as much as the second, and about seven and a half times as much as the first, was rated to be the cheapest by the Surveyor-General, considering the amount of information it could give.[215]

Just then, another controversy, almost similar, but different in nature, came up in the history of surveys—unequal cost of revenue surveys undertaken in different provinces and districts as well as the inconvenience and expenditure caused by the absence of a definite understanding between the Survey and Revenue Departments, as to the extent and character of the work which one department required from the other.[216]

This controversy cropped up in the North-Western Provinces where the Board of Revenue, found in 1871 that the settlement surveys of different periods gave inaccurate results. This forced the Board of Revenue to recommend the introduction of professional surveys by the Imperial Department. But after the new arrangement was introduced it was noted that the cost of cadastral surveys had gone up, i.e. there was not much difference between the cost of the earlier settlement surveys and the newly introduced cadastral surveys conducted by the professional department.

Another point which agitated the minds of the authorities was whether the work of the survey parties, employed upon topographical surveys, could be utilised for cadastral purposes, or vice-versa, i.e. maps which have been supplied by non-professional cadastral parties could likewise, be utilised for topographical purposes.

The Government of India, in consultation with the Surveyor-General, in a Committee meeting, held on 14 March 1882,[217] came to the conclusion that

> ... if a topographical party is working in a district which will sooner or later come under a cadastral survey, it will be desirable, if possible, to defer the topographical survey until the cadastral survey can be undertaken ... by professional agency. The cadastral maps can then be made to supply all the topographical information wanted. If, however, it is necessary for the Topographical Department to proceed with its work without waiting for the cadastral survey, it should be required at least to provide skeleton maps, which may be utilised hereafter for cadastral purposes. Whatever agents may be subsequently employed for field plotting they will only have to fill up the blocks of the skeleton maps supplied by the topographical party. What may be termed the scientific portion of the survey, will have been completed for them. The Government of India regrets that there has not hitherto been sufficient cooperation between the revenue and survey authorities to ensure the utilization of topographical work for the purposes of revenue survey in the manner indicated....[218]

Unfortunately, such cooperation between these two departments of revenue and survey authorities, in actual practice, did not exist. This was evident from the fact that the whole of the Central Provinces was surveyed topographically but it did not provide the revenue authorities with skeleton maps for them to use in cadastral or revenue surveys.

Of the Bombay surveys, the Surveyor-General maintained in regard to 'the extent to which Topographical Detail given in the *Settlement of Maps may be utilised in the Maps of the Topographical Surveys*'—that considerable differences of opinions still existed. No general ruling could be laid down because each Settlement Survey must be considered on its merits, and the ranges of merit in the surveys of different districts, and even of different portions of the same district were known to vary considerably. The topographical survey by this Department was mainly executed by plane-tabling on the basis of points fixed by triangulation or traversing. The accuracy with which different portions of the interior details were delineated, was very variable; it was greatest at the 'stations' at which the plane-table was set up by observations to the surrounding primary points, and at the subsidiary points subsequently fixed by rays from the plane-table stations, all of which were laid down mechanically with almost perfect exactitude; it was least in the areas lying between these stations and points, the delineation of which was not performed mechanically throughout but was mostly filled in by sketching, which required considerable judgement, and skill in estimating distances, on the part of the surveyor. The interior detail of settlement surveys was usually obtained by chaining, but without any framework. Thus, these surveys were liable to considerable accumulations of errors.

In the Bombay Revenue Surveys, the conditions were very different from those which were ordinarily met with in the settlement surveys of Bengal Presidency. In Bengal, permanent marks on the ground were rare excepting on the boundaries of villages, and even these were often missing; on the other hand, in Bombay the junction points of the fields were marked, and the marks were permanent, because these were preserved by the people of Bombay. This reality greatly facilitated combination with topographical survey, because of the numerous fixed points thus furnished which could be readily identified. This enabled the topographer to produce a far more valuable map of enduring usefulness by incorporating much of the work of the settlement into his own than he otherwise would be in a position to produce.

It was, therefore, with the object of enhancing the immediate value and future utility of the maps of the Gujarat Topographical Survey, rather than of reducing their cost, that the Surveyor-General at that time had endeavoured from the very beginning of that survey to have the utmost possible use made of the village maps of the earlier Revenue Surveys.[219]

So far as the Madras Surveys were concerned, all the surveys were carried out by the state department itself. No Imperial party was involved. In this Presidency the cadastral surveys had preceded the Topographical Survey, and, therefore, the ground had only to be gone over once. The Surveyor-General had experienced great difficulties in preparing Topographical Survey skeleton maps that would be utilised for cadastral purposes, without a large increase in the cost of topographical survey. The stations or marks of the Topographical Survey in the Madras Presidency had not largely disappeared; and the system of survey was somewhat different in this Presidency from that of the other Presidencies. In a nut-shell, it may be stated that the Madras Revenue Survey Department was really a Revenue and Topographical Survey Department combined; its official designation of 'Revenue Survey Department' was a misnomer and leads to misconception. It should have been called the 'Madras Survey Department'.[220]

We may now examine the progress of marine surveys in the delineation of the Indian coasts till the end of the nineteenth century, the take-off stage to the modern era. The marine surveys moved on in two directions—one on the land all along the coasts and the other surveys carried from the seaside aboard survey ships or vessels. The operations were almost similar in nature. Fixed points or important and fixed landmarks were observed; their astronomical positions were calculated and these were connected with trigonometrical survey stations while carrying the chains of small triangles all along the coasts, measuring and calculating distances between the points, taking into account the mean sea level. This, in brief, was the system of coastal surveys.

Other aspects of marine surveys like observations of soundings, marking of underwater dangerous rocks, etc., are of no significance here as what is being studied here is evolution of the topographical map of India and not the underwater topography of the Indian Ocean.

Contribution of marine surveys in the correct delineation of India's coastline, was at least as important as the share of the topographical surveys on coastal land. Both these were supplementary to each other. We have already seen the operations of the naval surveys in this direction till the end of the eighteenth century.

In the beginning of the nineteenth century, the Surveyor-General of India was equally responsible, for the marine surveys as for land surveys, and fulfilled the requirements of the Marine Board regarding the coastal charts.[221] In 1808, the Directors appointed John Wales as Marine Surveyor to the Supreme Government in India with his headquarters at Fort William and made him responsible for all marine surveys in eastern waters.[222] In 1809, the Marine Survey Department was established in Bengal, with John Wales as Marine Surveyor.[223] Wales however died on 15 January 1810 and was succeeded by Charles Court who reported his arrival on 13 February 1812. Charles Court held charge of the Marine Survey Department, Bengal, first as Marine Surveyor since 1812 and a few years later, as Marine Surveyor-General till he died on 9 September 1821.[224] He was, in turn, succeeded by Daniel Ross.[225] Ross resigned his appointment in November 1833 and was succeeded by his assistant, Robert Lloyd. He held the post till it was abolished in 1840.[226] The survey work however continued.

We have already noted that the coastal survey had already been completed by the beginning of the nineteenth century, though with inferior instruments. The survey work until 1828, entrusted to the Marine Department and carried out by Indian marine surveyors, was mostly concerned with regions far away from India, like the Red Sea, China Sea, etc.

The first marine survey of some importance on the coast of India was conducted by R. Cogan in 1828 on the coast near Bombay. The work on the coast between the latitude of the island of Bombay and the mouth of the Bancoot river was completed 1832. During 1828, Laccadive Islands were surveyed by Robert Moresby. To establish a navigable channel to the Bay of Bengal by the Pamban passage from the south side of the Adam's Bridge, survey work was taken up in 1837. But, because of the high surf at that season, the project was abandoned and the party moved westward and continued surveying along the coast of Madura. By the middle of 1838, the party was recalled. During 1837 and 1845, the following surveys with their charts were completed:

(1)	Pamban Passage	-	1837
(2)	Coast of Madura	-	1838
(3)	Western side of Palk Straits	-	1838
(4)	Island of Ramisseram and Manaar	-	1838
(5)	Palk Straits	-	1845

On the west coast of India, the marine surveys were carried out on the coast of Kathiawar and the Gulf of Cambay. The following charts were then prepared:

(1) Entrance of the Gulf of Cutch	-	1833
(2) Coast of Khathiawar from Diu of Dwarka	-	1833
(3) Diu Harbour	-	1833
(4) Kathiawar from Diu Head to Perim Island	-	1836
(5) Gulf of Cambay	-	1845

During this period the portion from the mouth of the Taptee, southward to Danoo, and the coast from Bassein to Bombay, were also surveyed.

In the north, in Bengal, river Hooghly was surveyed from Saugor Island to Calcutta in 1835; and during 1840, the sea face of the Sundarbans from Chittagong to Hidjellee was surveyed.

In July 1838, Sir Charles Malcolm was superseded as the Marine Surveyor by Robert Oliver. This proved disastrous for Indian marine surveys. The war like conditions in the Indian subcontinent, Persian Gulf, Afghanistan and China also adversely affected the survey activities, which were almost suddenly stopped. Oliver considered the Indian marine surveys useless and their prosecution was considered by him as 'waste of public time and money'. These views also greatly contributed to the sorry state of surveys during his time.[227]

With the easing of situation during the time of the successors of Robert Oliver, from 1848 onwards, the survey activities again picked up. During 1848-50, the surveys of Lacadive Archipelago, and the entire approach to Bombay harbour were completed; the marine surveys of North Canara were connected with those of the Guzerat Coast.[228]

In October 1850, A.D. Taylor was appointed as the Marine Surveyor and he completed the survey of the Gulf of Cutch and the coast of Kathiawar in that month. In November 1850, the coast from Cape Comorin to Beypore was trigonometrically surveyed. Next year, the North Concan coast was surveyed, along with Choul harbour; the soundings off the Bombay harbour were also observed. In the year 1852-3, the coastal surveys were started from the mouth of the river Tapti, went up the coast of Kathiawar in its northwest region finally up to Vomani Point.

Even after the abolition of the Office of the Marine Surveyor in Bengal in 1840, marine surveys continued and between 1841-8, detached marine surveys were continued on the Coromandel coast by W. Fell. During 1851-9, the Bassein river survey was concluded.

In 1860, the remaining small portion of the Malabar coast was also concluded. This completed the entire western coast of India during this century.

In the east Indian Ocean, W. Marshall concluded the survey of Port Blair, Middle Straits and the Cocos Islands during 1859-60.[229]

With the abolition of the Indian Navy in 1862, no arrangement was made to continue the survey operations. It was in 1875 that a small department was again organised under the superintendence of A.D. Taylor, who had a good personal record of marine Surveys.[230]

Earlier in 1861 India Office had transferred all the materials available with them for the construction of sea charts, to the Admiralty. From 1861 till 1871, a period of ten years, little or nothing had been done in the field of marine survey except some desultory operations, without any combined plan, and with inadequate means and '... in most cases the work was left in the hands of persons who were quite unacquainted with nautical surveying except in the rudest form'.[231]

During the last fifty years, extraordinary changes had taken place in the configuration of the coast in many areas, as the last surveys executed were some 30 to 50 years old.

> ... Ports which were of no importance then, and the examination of which consequently had been but cursory, had now become open to commerce, and yet the charts remained the same and were practically useless ... surveys of half a century, or more ago, were merely preliminary examinations, and could not, in the first instance, without ... trained officers, have been executed in the precise manner in which surveys are now carried out by the Surveyors of the Royal Navy, the rigorous exactitude of which is absolutely necessary for the requirements of the present advanced state of navigation and commerce.[232]

After receiving representations etc. from various personalities interested in the Indian marine surveys, like a Memoir from Clement Markham, memoranda from T. Saunders, Geographer to the India Office, Constable of Indian Navy and A.D. Taylor, the secretary of State for India, in 1871, under his Despatch No. 7 (Geographical), dt. March 30, 1871, India Office, forwarded these memorandums and the Memoir and brought the subject of the sorry state of Indian marine surveys and the related problems to the notice of the Government of India. Instead of giving their opinion, the Government requested the Court of Directors to depute Commander A.D. Taylor to India to assist them in devising suitable measures to remedy the inconveniences pointed out.

This was approved, and Taylor reached Calcutta in December 1873. He examined all the existing charts along with the material for charts available in India, and in England of the entire coast from Pakchan Estuary at the southern extremity of Tenasserim, in Burma, to Sonmiani Bay on the western limits of the Sind coast, now in Pakistan, including the islands in the Bay of Bengal and worked out a scheme to supplement and perfect the êxisting charts, the details of the agency by which it should be carried out, and its probable cost. The marine authorities, and others interested in the coasting trade and approaches to the Indian coast and harbours, were also consulted. A report was then submitted to the Secretary of State for India.[233]

The proposals submitted thereunder, were approved by the Secretary of State for India.[234] These were summed up as under:

> ...Completion and maintenance of the charts of the Indian coasts are urgently needed....
>
> ... As regards the cost of marine surveys, you (Government of India) propose to take responsibility for all the coasts of British India,... with certain groups of islands....
>
> ... You propose to adopt measures which will secure the annual execution of a satisfactory amount of surveying work, commencing with those portions of the coast which most urgently require examination ... the general control will be entrusted to a Superintendent of Marine Surveys....[235]

It was also agreed that the charts of the Indian surveys would be printed and published under the superintendence of the Hydrographer, at the expense of the Admiralty.

Great importance was attached to the work of connecting sea surveys with shore surveys by means of the points of the Great Trigonometrical Surveys.

The records, charts, etc., that were transferred to the Admiralty in 1861, were placed at the disposal of A.D. Taylor who was appointed Superintendent of the Marine Surveys in April 1875. A small office was established for him at Calcutta. The Department thus created, commenced work '... with those portions of the coast which most urgently require examination....'[236] The examination of the entire coast, harbours and their approaches, etc., was undertaken; decisions regarding priority of surveys and re-surveys wherever needed, were taken in right earnest by A.D. Taylor.

After initial examination, the work on the Indian coasts (excluding Burmese coast) began in 1880. From 1880 till 1890, the entire coast of India was surveyed and re-surveyed. All the ports, including places of anchorage, from False Point on the Orissa coast to the Bombay harbour, were examined and properly surveyed.

Taylor had surveyed the Concan coast below Bombay from 1843 till 1847. From 1850 to 1855, the Gulf of Cutch and the Malabar coast were surveyed; from 1855 to 1859, Karwar, Coringa Bay, Coconada port on the Coromandel coast, and Cochin on the Malabar coast were also surveyed. As stated earlier, the marine surveys were stopped from 1862 onwards. Taylor remained in office till September 1881.[237]

The following ports were visited and examined:[238]

1.	False Point	11.	Colachel
2.	Gapalpur	12.	Cochin
3.	Calingapatam	13.	Narakal
4.	Bimlipatam	14.	Beypur
5.	Vishakhapatnam	15.	Calicut
6.	Coconada (Calinga)	16.	Tellicherri
7.	Masulipatam	17.	Cannanur
8.	Madras	18.	Mangalore
9.	Negapatam	19.	Karwar
10.	Tuticorin	20.	Bombay

These marine surveys ultimately helped in the correct delineation of the entire coast of India and putting the topography of India in a correct perspective, which favourably affected the position and standing of the survey of India amongst the world community of cartographers, as well as helped to usher India in the modern age.

In the meanwhile, on 6 October 1882, the following review of the Annual Report of the operations of the Survey of India for 1879-80 appeared in the *London Times* a daily newspaper. This created wrong impression in the minds of the public in England as well as among the authorities there. The review had commenced with the remarks

> From the just received report of the operations of the Survey of India for 1879-80, we learn that this stupendous work is now all but completed.[239] The Surveyor-General J.T. Walker, however clarified from Calcutta that—
>
> "It is not however, the General Survey, but the Principal Triangulation, which is stated in the report to be approaching completion. The Principal Triangulation has come to be widely known, out of, as well as in, India, as it was executed with the object of serving for geodetic and other scientific requirements, conjointly with similar operations in Europe, in addition to fulfilling its primary object of furnishing a basis for all survey operations in India; thus it is occasionally, as in the present instance, confounded with the General survey of which it is in reality but a part, though very important part, and one which must necessarily be completed long before the whole can be completed. The principal triangulation of Great Britain and Ireland was finished about thirty years ago, whereas the general topography is still far from finished. Thus in India the approaching completion of the principal triangulation does not imply that the general survey is also approaching completion."[240]

The Surveyor-General of India, in August 1901, wrote to the Government of India, intimating to the authorities the great necessity for making arrangements for keeping up todate the modern topographical maps of India which have lately been issued or were at the time in the course of survey and compilation.

He also urged upon the authorities the urgent need for an agency for revising the 'standard sheets' by going over at stated period of intervals, and all small changes, either caused by nature or by the hand of man entered therein. He suggested a period of ten or fifteen years for the said revision. He informed the Government that in India, the 'standard sheets' were completed on a 1-inch scale and this scale had been adopted as the scale on which a general map of the country, required for purposes of efficient administration, both civil and military, was to be compiled and brought out.

But the problem was that no sooner was any area of the country surveyed than the map, representing that area began to get out of date. The process involved was of varying rapidity. Many factors were responsible: new roads, canals and railways were being constructed; village and other boundaries were constantly being altered; and, most difficult of all to cope with were the rivers, especially those flowing through alluvial tracts which changed their courses in the most bewildering way.

The Surveyor-General also lamented that in India up to that time nothing at all had been done by way of real revision of the then existing maps. The result of this neglect of revision was that the maps, which were at one time fully sufficient for every need of the administration, were now hopelessly out of date and of little or no value.

So J.T. Walker emphasized the need of establishing an agency in every province for revising the maps constantly so that they might not fall too far into arrears. A sound system of collecting information, as to changes that had taken place during the year, was also suggested.

For all the problems posed by him, a reorganisation of the priorities of the Department along with changes in their strength, training and the functions, were also proposed by the Surveyor-General.

To facilitate a proper study of the problems a 'Note on the present state of the Topographical Maps of India' by the Surveyor-General of India, which has been discussed earlier, was also appended to his letter to the Government of India.[241]

In their reply, the Government expressed their complete concurrence with the proposal that arrangements should be made for the correction of 1-inch topographical sheets which had become obsolete. They said

> ... so far as the necessary changes merely entail the plotting of new roads, railways or canals, the scheme which you propose of working on information supplied by District officers.... District officers would give lists of the features of this class which are missing on the map sheets, and surveyors would be deputed to plot them. It would not ... be necessary to strengthen the Provincial staff of officers for this purpose.
>
> In a country which is not liable to alluvial action, very few changes can occur in the physical features indicated on the map sheets, and it is believed that where revision is required it is rather to correct inaccuracies and to complete omission in the original surveys and mapping than by reason of changes which have rendered the maps obsolete. The Government of India agree that it is desirable that the map sheets should be maintained on a uniform standard, and that revision should be undertaken where it is needed for the purpose.... It is obvious that for revisions of this class reliance cannot be placed on information obtained from District officers. A general examination of the country will be necessary.
>
> Where, as in Bengal, the shifting of rivers and streams materially change the features of the country, periodical revision will be necessary. But in this case again it will be impossible to rely upon information collected in the Revenue Department otherwise than as a very general indication of the extent of revision required, and it will be necessary to deal with each district separately and institute a general comparison of the map sheets with the ground. It appears that shiftings of water-ways have rendered obsolete a good deal of the mapping of Lower Bengal. But it is understood that you do not contemplate a revision of the map sheets in anticipation of the extension of the cadastral survey.
>
> As regards the maintenance of village triangulations, their demarcation is carefully protected by the Land Record Staff in temporarily settled Provinces. In Behar there is less security against their removal, as no Land Record Staff has as yet been established. But the maintenance of the cadastral maps and records is now under the consideration of the Government of India, ...[242]

In July 1904, while reporting on the General Report of the Operations of the Survey of India, during 1902-3, the Government of India indicated that a Committee would be appointed to report upon the working of the Survey of India.[243] Circulars in this regard, were issued to all Local Governments and Administrations by the Government as early as 7 July 1904.[244] The Committee was constituted 'to consider the condition of Topographical Maps and the measures to bring them up to date'. The Preliminary Report of the Committee, along with the comments of the Department of Revenue and Agriculture, Land Surveys, which obviously included the opinion of the Surveyor-General of India was sent to the Secretary of State for India in September 1904.[245]

The Governor-General in Council decided, on 31 October 1904 with the approval of the Secretary of State for India, to appoint yet another Committee under Resolution No.1581-70-20 for the purpose of making a complete examination of the then existing methods and working of the Survey of India Department in all its branches, with special reference to the preparation or revision and the reproduction of the topographical maps of the country.[246] The Committee was called the Indian Survey Committee which finally met in November-January 1904-5.

The more important points, among others, which the Committee would enquire into, and report on, were the following:

1. The state of the topographical maps in each province and the measures required to bring them up to date
2. Methods and expense of survey
3. Methods of reproduction
4. Organisation of the Department.

The views of the Local Governments were also be sought on these points.

Apart from other recommendations of the Indian Survey Committee, the one that was more important and relevant here was that the topographical surveys of the greater part of the country had indeed been brought to completion many years ago, but they were conducted on no uniform principle, and even where the surveys were originally accurate, they had in many cases become obsolete by mere lapse of time.[247]

The Committee also proposed for the completion of the surveys of parts of the country of which no maps existed and of which the existing maps were defective; and for the revision of all maps at intervals of 15 years.

The Department was aware of this but because of shortage of staff and funds, nothing could be done by way of bringing old maps up to date. The demands made on the staff and funds available with the Department adversely affected new surveys of tracts or countries which had been annexed, or the advancing frontiers, and professional cadastral surveys in Bengal and elsewhere. Even the policy of the Government of India in the past had not been such as to encourage proposals likely to involve a large expenditure on work that was not directly remunerative.

The result was the preparation of a wider and more comprehensive scheme than that which was put forward by the former departmental committee. The Government of India agreed with that as it had proposed a definite programme on clear-cut lines, employing survey parties to work systematically in well-defined areas and by concentrating the whole strength of the Department, as far as was possible, on its topographical work so that the map of India could be brought up to date in a satisfactory and business-like manner. In other words, the preparation of a modern map of India, on a 1-inch scale would at once be undertaken according to a definite programme. There would be, in all, fifteen topographical field survey parties employed for the purpose. The work of the preparation of this modern map of India, as per their calculations, was to be completed in twenty-five years. This would mean that there would be a regular revision of the maps after every twenty-five years. The scale of the survey would generally be at least double that of publication, i.e. it would be 2-inch to a mile.

The second recommendation was that the improved process of heliozincography would be employed to print maps, but a few selected standard sheets could be engraved on copper.

Third, the Atlas of India sheets should be superseded by 'degree' sheets on the $\frac{1}{4}$ inch scale, which would be engraved on copper. So, the publication of the Atlas of India sheets would, henceforth, be stopped.

Fourth, publication of $\frac{1}{16}$ inch provincial maps would also be discontinued and be replaced by a general map of India on the international scale of 1:1,000,000, which would also be engraved.

Fifth, the earlier cadastral and other large-scale surveys would be left entirely under the control of Local Governments, who would provide for the expense of the work and also for the production of the maps.

Sixth, village boundaries would not be shown on the ordinary topographical maps but if these were required for administrative purposes, a special edition showing the boundaries would be issued under certain conditions.

Seventh, the Survey of India would give up the expensive and scattered surveys by the Imperial parties of forest areas and if considered necessary, to arrange for surveys of such forest tracts as had not already been dealt with on a 2-inch scale, and the cost of special forest surveys would be debited to the forest department.

Apart from the above mentioned technical recommendations, the Survey department was also re-organised to cope up with the new responsibilities. The staff of the Imperial Service for trigonometrical and topographical work was increased from forty to seventy officers; in place of the then existing Provincial Service, a new Provincial Service was recommended to be established on a better pay-scale and, another service formed intermediate between this and the present subordinate service, and more Indians were also to be appointed to this service.[248]

The Government of India had agreed to modify the scales for publication of maps wherever special requirements of the army persisted. The recommendations of the Army Council, as laid down by them in their memorandum submitted in connection with the Indian Survey Committee's proposals and recommendations were considered and agreed to by the Survey of India.[249] But in general, we may say that the recommendations of the Survey Committee almost remodelled the quality of Indian maps which stood the Survey of India in good stead in the International conferences and committees in future.

The standard scale of 1 inch to a mile was first introduced by the Survey of India in their survey programme for the year 1908-9. While reporting the progress, they stated

... The whole topographical programme is a continuation of last year's, the only change that has been introduced is that the 1"= 1 mile has now been adopted as the standard scale of survey except in the case of forests and of those special areas for which the military authorities require a larger scale.

No desert areas, for which the $\frac{1}{2}$ inch scale has been laid down as sufficient, enter into the programme for 1908-9....[250]

It were these one-inch-sheets from which most of the other Indian mapping was derived later. The initial sheets were published in 5 colours, red for roads and sites, blue for water, brown for contours, green for wooded and jungle areas, and black for other details and lettering. In the year 1911, a sixth colour, yellow, representing cultivated areas, was introduced, and later, a seventh, grey, for half-tone hill shading was added. A distinction was also made between water forms that normally hold water and those that were generally dry. Dry water areas were being marked with black. The roads were classified under six heads according to their suitability for different forms of traffic. Approximate contours were shown at 50 feet vertical intervals, wherever possible. The first sheets appeared during 1907.

We may now describe other standard or normal series of maps that were published.

The one-inch scale was first intended to cover the whole of India, but for the less developed areas of the country, the half-inch scale was later found to be sufficiently large. It was also found to be the most suitable for general military purposes. The whole of India, including those areas which had been already mapped in the modern one-inch scale, was then decided to be mapped on the half-inch scale. The modern half-inch maps had already been following the general design of the modern one-inch maps in style and colouring.

The Degree sheets, or new quarter-inch maps, replacing the old Atlas sheets, on the same scale, were compiled by photographic reductions of one-inch or half-inch sheets, and were published by heliozincograph process. These were, obviously, reduced by photography in the same style and colouring as the one-inch sheets but, being reduced, had to be more generalised. For example, unlike on Degree Sheets, no distinction was made between cultivated and non-cultivated areas; the hills were shown by contours, but generally at 250' vertical intervals, instead of 50' intervals. These were supplemented by hypsometrical layers and hill shading.

Another series that was initiated was called the 'India and Adjacent Countries' series. This was on 1:1 million scale. These sheets were, in the beginning, made from the old topographical maps which did not show contours but used hechures to represent hill features.

From the 1:1 million sheets, another series was made —each new sheet was comprised 12° of longitude X 8° of latitude, the scale being 1/2 million, i.e. about 32 miles to 1-inch. This began in about 1909. This series was called 'Southern Asia' series. Again, the style and colouring remained exactly the same as that of 1: million scale series.

Till the beginning of the above mentioned series of modern maps during the first decade of the twentieth century, for the greater part of India, old lithographed or photoprinted one-inch maps, the old standard sheets, including 30' of longitude and 15' of latitude, without contours and hills shown by hechures or form-lines, printed mostly in black and occasionally in black and brown, were available and made use of.

Gradually these map sheets were superseded by contoured sheets of half their size, from modern surveys, printed in 5 colours including black. For the more highly developed regions of the country, these modern maps were published in two series, one on the 1-inch scale and the other, on $\frac{1}{2}$-inch scale. Along with the newly published, post-1905 maps on modern scales, there was another complete series of engraved quarter-inch Atlas of India maps available. These were printed in black, with hills shown by hechures. Though much in demand, they were gradually superseded by Degree Sheets of contoured and layered quarter-inch coloured sheets.

Apart from these, there were many categories of geographical general maps on various scales being published by the Government for general use.

One of the earliest of the Survey of India general maps of the country was a map on the scale of 32 miles to 1-inch. This was chiefly compiled from Trigonometrical Surveys, and engraved under the orders of the Directors of the East India Company, by Mr. John Walker in 1852. It was followed by many other maps on smaller scales. These were mostly engraved and published in England by the orders of the India Office.

An engraved map of India, published by the Survey of India, on the scale of 32 miles to 1-inch, was published in 1881, and was entitled *The Map of India and Adjacent Countries, scale 32 miles to one-inch, 1881.* The topography was drawn in mezzotint chalk shading on stone. This map went into four editions and the last edition was printed in 1908. The map was engraved in 6 sheets. This was superseded by another map, with the same title and scale, and compiled from the old topographical maps, engraved in 12 plates and its first or 'geographical' edition, was published in 1913 with hypsometrical layer colouring.

The compilation of such a map from the old topographical maps covering an area bounded by 40° X 30° of the earth's surface on 32-mile scale, was a difficult task. There was no complete data, in many cases, even no precedence in Indian cartography was available at all. The only data available for contours, except in a few isolated areas where modern surveys were in progress, was the spot heights on the topographical maps. By interpolating the contours from them, a fair approximation to accuracy was obtained. As for a wall map of the size under compilation, the ten land and three bathymetrical layers mentioned in Table 8 were adopted as the minimum that would show the general formation of the area:

TABLE 8

Land Layers		
0	—	500 feet
500	—	1,000 feet
1,000	—	2,000 feet
2,000	—	3,000 feet
3,000	—	6,000 feet
6,000	—	10,000 feet
10,000	—	15,000 feet
15,000	—	20,000 feet
20,000	—	25,000 feet
Above	—	25,000 feet
Bathymetrical Layers		
0	—	5,000 feet
5,000	—	10,000 feet
Below	—	10,000 feet

Thus, the wall map was compiled in such a way that, in the first place, the physical geography of the whole country could be seen at a glance—boldness in drawing, lettering and layer colouring were maintained and minor details were either omitted or generalised, as necessitated by the scale.

In 1916, a physical map of India, or the political edition of the new 32-mile map was published, which was followed by a layered and shaded edition in 1917. As per the decision of the Government of India taken in 1913, a distinctive colour was applied to the boundaries of each province and country, and the colours decided upon were yellow for native states, dark red for British India, and light-red for the regions under British Administration.[251]

In December 1905, after knowing the views of the Indian Survey Committee (1904-5), the Survey of India sent for the approval of the Government of India, a scheme for numbering the sheets of the Standard Series of maps of India and its adjacent countries. This Survey of India also decided to reduce the size of 1-inch Standard sheets to one-half, viz., from 15° latitude X 30° longitude to 15 minutes of latitude and longitude, and also to replace the Atlas of India series by Degree Sheet series by the scale of one-inch to four miles, which was finally modified by the Indian Survey Committee to 1:1 million scale, and the Degree Sheet to be of 4°X 4° in size, as will be explained later.[252]

The scale on which the general maps was to be surveyed and published was one of the most important matters that was dealt with by the Indian Survey

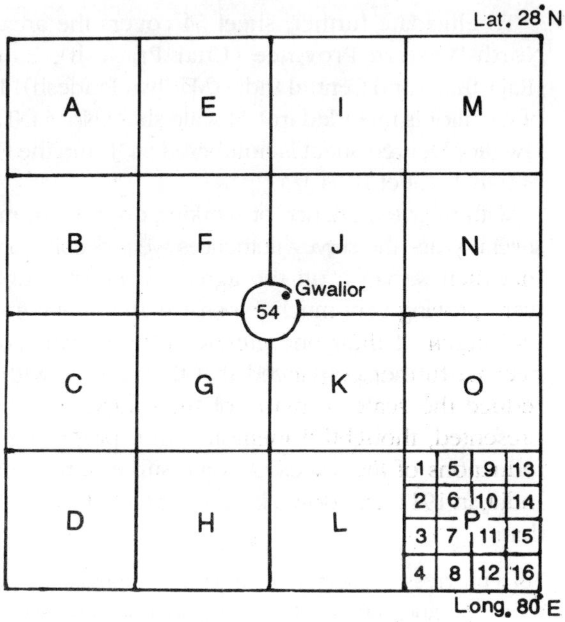

East Sheet No. 54, Scale 1/Million
Long Gwalior-54 J, Scale 1-4 miles
Gwalior-54 J/4, Scale 1-1 mile

Committee while laying down the programme for future operations: The Committee recommended

> That the scale of survey should ordinarily be 2-inches to the mile, and the scale of publication 1 inch. It also proposed for smaller scale maps to substitute degree sheets on the scale of 4 miles to the inch for the existing atlas sheets which are on approximately the same scale, and to issue a series of sheets on the scale of 1/1,000,000 (1/M), or approximately 16 miles to the inch. Each degree sheet would be compiled from 16 1-inch sheets, and each 1/M sheet from 16 degree sheets. The 1-inch sheets would represent 15 minutes of latitude by 15 minutes of longitude, and by this arrangement every 1-inch map would be an integral part of one of the degree sheets, and every degree sheet an integral part of the one of the 1/M sheets.[253]

The above could well be understood as follows:

The recommendation of the Indian Survey Committee initiated what may be called the modern topographical survey of the country in 1905. This was a very systematic scheme for the Indian mapping. Under this scheme the whole of southern Asia between the meridians 44° and 124°, east of Greenwich, and between parallels of 4° and 40°, north latitude, was divided up into sheets of 4° latitude X 4° longitude, on the scale of 1:1,000,000, the sheets being numbered consecutively from the north-west. Each Indian 1:1 Million sheet was to be further subdivided, into 16 sheets of 1° latitude X 1° longitude, on the scale of 1 inch to 4 miles, known as 'Degree' sheets and lettered from A to P. Each Degree Sheet was, in its turn, subdivided into 16 one-inch to one mile sheets of 15' latitude by 15' longitude numbered from 1 to 16.

To elucidate further, sheet 54 covers the area of the North-Western Province (Uttar Pradesh), Rajputana (Rajasthan) and Central India (Madhya Pradesh). The city of Gwalior is included in 1:M scale sheet No. 54. Thus the Gwalior Degree Sheet is numbered 54 'J' and the Gwalior one-inch sheet is 54 J/4.[254]

With their experience of working on modern maps for several years, the survey authorities were deeply convinced that their surveys, on the agreed scale of 1:1,000,000, were proving very much expensive and time-consuming, and much of their energies were being wasted. They became further convinced that they could, with profit, reduce the scale of many of their surveys. In a well-presented, thoughtful memorandum, proposing certain reductions of the scales of some surveys, the Survey of India, in 1911, laid down before the Government of India that:

For a progressive country like India, it is impossible to lay down a final programme of survey; the only practicable course is to consider the present requirements of the various regions and to decide on the scale for maps most suitable to each region.

Our new $\frac{1}{4}$-inch maps are in their infancy; there is room for improvement in them, and they may be made to show a larger amount of useful information than they do at present, without being over-crowded.

The plan outlined in this memorandum will give a $\frac{1}{4}$"- m map to the whole of India.

It will also give a 1" = 1m map to the whole of British India with the exception of the 282,000 sq. miles described in para 3. [Reduction of scale to $\frac{1}{2}$ of Bellary etc., C.P. and Chota Nagpur, wild country between Assam and Burma,.... and reduction of scale to $\frac{1}{4}$ — of Higher Himalayas and Baluchistan]....

It will also give $\frac{1}{2}$"=1 m. map to the areas of the Native States, and it will enable the major Native states to say whether they desire the 1"=1 m scale or not....[255]

A circular, to this effect, was issued by the Government of India to all State Governments and Administrators in February 1912

In para 3 of ... Circular No. 3, dt. 11th January 1907, it was stated that the Governor-General in Council had accepted the recommendations made by the Survey Committee of 1904-5 that the scale of topographical survey should ordinarily be 2"= 1 m and that the maps based on this survey should ordinarily be published on the scale of 1"=1 m.

It was subsequently, however, decided ... the scale both of survey and of publication should ordinarily be that of 1"=1 m, except (1) in special cases, where a larger scale was necessary, and (2) in underdeveloped areas such as, the greater part of ... the Rajputana Desert where the scale both of survey and publication would be reduced to $\frac{1}{2}$"=1 m.

... The Surveyor-General has since proposed to the Government of India the area in which a scale of less than 1"=1m shall be adopted for survey and publication, should be considerably increased ....

... whether there is any objection on the part of Local Governments... [to the above proposals].[256]

The Government of India ultimately during 1912, agreed and reduced the scale under which all Native States and certain parts of Eastern Ghats of southern India, were to be surveyed on the $\frac{1}{2}$-inch and not on the 1-inch scale. The area affected by this decision was approximately one-third of the whole of India.[257]

At the time when the Survey of India was burdened with the preparation of maps of two different series, 1:1 M and $\frac{1}{4}$-inch scales, the detailed specifications, including scales, used on Indian maps, came under prolonged discussions, covering over several agencies, including the Army Council.

The question of the most suitable scale for the tactical map of India was being discussed by the military authorities for quite some time. The Survey Committee of 1904-5 had recommended the scale of 1 inch to one mile for the purpose. The Army Council in London had pointed out in 1907[258] that the 1-inch scale was too large for military purposes and had opted for the $\frac{1}{2}$-inch scale. The Government of India was in a fix. For the last many years, topographical maps of India had been published on two scales, namely 1-inch and $\frac{1}{4}$-inch scales, the standard sheets and Atlas sheets respectively. The Survey Committee had recommended the retention of these scales. The maps on these scales were in general use amongst Civil officers and were well understood by the public.

Fortunately for the Survey of India, the Army Council's recommendations for $\frac{1}{2}$-inch was turned down by the Army Department, which accepted the scale of the old Indian Atlas sheets, i.e. $\frac{1}{4}$-inch scale, for their tactical maps, even though there were some drawbacks.

During 1914, the Army Department revised its earlier decisions and had come to the conclusion that neither of the Civil maps of India, namely on the 1-inch scale and $\frac{1}{4}$-inch scale, were suitable for military purposes. They were of the opinion that the 1-inch scale was too large, that the $\frac{1}{4}$-inch scale was too small, and that the adoption of the $\frac{1}{2}$-inch scale for the tactical map had become a military necessity. The adoption of the $\frac{1}{2}$-inch scale for the military maps, as proposed, would require, as calculated by them, lesser number of topographical sheets than what was proposed by the Survey Committee or by the Army Council earlier.[259]

The Government of India agreed to the proposal[260] of $\frac{1}{2}$-inch scale for tactical maps and the necessary instructions were issued to the Survey of India.[261] The Government of India, in their letter also requested the Survey of India that they be informed of the probable duration of the topographical programme under the altered conditions.

The Survey of India could not give a clear-cut answer to the query. They stated that the additional new assignment had come at a time when the Department was passing through a situation created by the First World War of 1914-19. Because of this war, field programmes were curtailed during 1914-15 and would again be further curtailed during 1915-16; many surveyors were employed in the reduction of 1-inch maps to $\frac{1}{2}$-inch maps from the vast collection of such 1-inch old surveys of the past 10 years, instead of their being employed in the field surveys. This programme, they replied, would probably be completed by the end of 1916. In the meantime, topographical outturn would be considerably reduced during 1914-16. Even after 1916, some of the surveyors would continue to be employed on 1-inch maps from the new 1-inch surveys. All these would affect the ultimate programme of completion of topographical mapping from 1935 to perhaps 1940. These forecasts were also likely to be vitiated by other unforeseen causes that might arise in the coming next 25 years, like reduction of budget allotments, transfrontier campaigns, territorial additions, further surveys in Assam and Burma, etc.

Even military requirements, they added, might be changed as the existing maps were designed in the past and the need for new sizes, symbols, contours, colouring, scales, etc., might as well alter the situation, with the new experience being gained in the war that was then going on. The new tendency in Europe was in favour of increase in the scale of maps for military purposes to such large extents that the introduction of the new $\frac{1}{2}$-inch map for India, in place of the $\frac{1}{4}$-inch as the tactical map, might prove to be a small scale as compared to those additional requirements being felt for artillery and even for aeroplanes in the foreseeable future.[262]

In 1913, the Government of India reviewed the recommendation of the Indian Survey Committee regarding the time taken to complete one cycle of the preparation of the 1-inch survey maps of standard sheets of the country and said that the time-table could not be adhered to and would now be completed most probably by 1945. For completing it by 1935, the original time schedule, they decided to reduce the scale of survey and publication to $\frac{1}{2}$-inch and $\frac{1}{4}$-inch respectively in certain areas where the population was not dense.[263]

Due to the addition of the preparation of the tactical map of India to the programme of the topographical mapping of the Survey of India, as now proposed by the Army Department in 1914 and agreed to by the Government of India, the Government intimated the Secretary of State for India[264] that in the latest estimation of the Survey of India, because of the new addition of $\frac{1}{2}$-inch map, after its earlier review of 1913, the cycle of completion of the 1-inch mapping within twenty-five years, proposed by the Survey Committee, would be disturbed, and the programme would prolong by about five years, i.e. until 1939 or 1940. But before this could be achieved, the Second World War intervened.

Several circumstances had operated against the concentration by the Survey of India of its entire strength, as agreed to earlier in the wake of the decisions of the Survey Committee on topographical work. The Department was called upon to prepare forest maps, greatly in excess of what was agreed to and also on a larger scale.[265] Even an item of work which the Survey Committee, had specifically, prohibited the Survey of India, to include in the Department's work, i.e. surveys of Cantonments, the Survey Department was asked to devote its time, energy and money[266] to it. The Department had also to initiate a City Survey of Bombay;[267] it had to assist the Punjab Government in their irrigation schemes;[268] and to re-survey the areas covered by certain Artillery Practice Camps for the Army Department. All these additional items of work were forced upon the Survey of India at a time when, as a result of the 1914-19 World War, the active staff was considerably reduced because of the reversion of most of the survey officers back to military duty and fresh formation of survey parties to survey in Persia. All these factors had forced the Department to curtail the annual field programmes of the topographical and trigonometrical parties since 1914.[269] In 1917, the Survey of India again assessed the situation and concluded that, having regard to various diversions and deviations, the topographical programmes could in no case, be completed before 1945. The Department was apprehensive of the post-war military requirements also as in the pre-war cases of the survey of town areas,[270] which might come up again.

It was felt in 1918 that time had come to review the position and to consider whether the organisation and the sanctioned programme of the Survey Department were suitable for the requirements of the near future.[271] The entire issue was discussed at a meeting held at Dehra Dun on 29 March 1918 when high-level officials deputed by the Secretary of State for India and the Indian officials from the Government and the Survey of India, attended. After the discussions, the Survey of India submitted concrete proposals[272] to the Government of India.

In the meanwhile, in the International Geodetic Conference (1909), proposals were made to link up the Indian Triangulation with that of Russian Turkestan for the purpose of carrying out Geodetic Triangulation over the Hindu Kush mountains.[273] Ultimately, the Indian triangulation was linked up with the Russian triangulation through the terminal stations near the Kilik and Mintaka Passes on the Indian side with Taghramansu and Beyik Passes, the terminals of the Russain triangulation respectively. The link-up completed the triangulation from Ireland on the west to Cape Comorin in the south and to Burma on the east.

It may be noted that the reorganisation of 1905-8 had made no provision for any series of maps on scales smaller than 1:1 million. The experience, that was acquired in the compilations of the 32-mile maps, had shown the practical difficulty of preparing a map of the whole of India in one operation that would include all the latest information available at any one time.

Towards the end of the nineteenth century, an international movement was gaining ground among the geographers in favour of the preparation of an International Map of the World in a uniform style and on one scale. The proposal was first mooted at the Fifth International Geographical Congress held at Berne in 1891. This proposal was discussed again and again in subsequent congresses. Eventually, an International Committee of cartographers and geographers, which assembled in London, under the auspices of the British Government, decided that a map of the world on the scale of 1:1 million would be produced, and that respective governments would prepare and publish the sheets of the map in which their territories were included, in accordance with the rules and specifications formulated by the Committee. Though India was not directly invited to any of these International Conferences till then, it was indirectly represented by the British Government.

In February 1911, the Director of Military Operations, War Office, London, had informed the Under Secretary of State for India

> the considerable amount of mapping on this (1:1 M) standard scale already carried out by the Survey of India was recognised and appreciated by the Committee, and it is now suggested that this work should be brought into exact conformity with the system laid down in the International agreement....[274]

This letter was passed on to the India Office, which in reply to this, forwarded a copy of earlier letter from the Survey of India addressed to the Government of India,[275] along with their reply to the Surveyor-General,[276] and went on to say

> ... the 1:1 M map now in course of preparation in India (Some 29 sheets of which have already been published) is part of a systematic plan which has been in operation for some years. In these circumstances it appears to be matter for regret that the Government of India had no opportunity of expressing their views on the subject at the meetings of the International Map Committee; and it is possible that they may find some difficulty in adjusting their system to the plans of the Committee unless strong reasons for doing so are laid before them. If, therefore, the War Office desires to press this change upon the Government of India for military reasons it seems necessary that this Office should be informed officially on the nature of these reasons and the weight attached to them by the Army Council, before any recommendation is made to the Government of India.[277]

The Director of Military Operations, War Office in their reply stated that the 1:1 M map, which was under compilation and publication in India, was well-suited for its purpose of a general and political map. He went on to mention

> ... The two maps need in no way conflict with each other. Up to the present, countries comprising over three-quarters of the entire land surface of the globe have announced their adhesion to the Resolutions of the International Map Committee. It would, for all the countries concerned, be a matter of great regret if the Government of India were unable to join in this International Agreement.[278]

They further clarified that the International Map was not based on military considerations.

The Government of India, in the Department of Revenue and Agriculture, (Land Surveys), accepted the proposals of the Director, Military Operations and directed the Survey of India Department to undertake the preparation of the International Map of India, on the scale of 1:1 M in accordance with the system laid down by the International Map Committee. While conveying their acceptance to the Secretary of State for India, they expressed the view that

> ... This decision will not affect the preparation of the standard map of India on this scale which is already in hand...[279]

The Survey of India had already advanced much in the preparation of their own 1:1 Million map sheets. They could not, therefore, alter their format without upsetting the whole of Indian cartographical framework, to suit the new international requirements.

These two maps, though on similar scales, differed in many ways. The two very important particulars were: size of the sheet and unit of height measurement. Each international sheet covered 6° of longitude and 4° of latitude, while the Indian sheets included 4° of longitude and 4° of latitude; and the time had not yet come when height could be shown in metres on maps to be used in India. There were other differences too: the Indian sheets used the foot, instead of the metre used on the International sheets, as the unit of height measurements. The lettering of the 'International' map was to be uniformly upright with the exception of that of water forms and communication names, while on the Indian map all the less important place names were in italics. The contours, by the decisions taken at the International Committee of 1913, were to be fine, black, dotted lines on the International map, whereas on Indian map, they were in brown continuous lines. The classification of site symbols was to be much more elaborate on the International map, than on the Indian map. On the International map, trees were to be shown in plan by small irregular circles, while on Indian map, these were not shown at all.[280]

Further, the Indian sheets could not be altered to agree with the International sheets also because they formed the key to the whole system of numbering and arrangement of the Indian topographical maps.

It became necessary, therefore, for the Survey of India to prepare two separate series of maps on 1:1 Million scale — one, their own, and the other 'International', as a special series. This was in addition to, and to be taken up after the production of the corresponding Indian sheets. The general impression of the authorities at that time was that by participating in the International task, India would definitely stand to gain, though indirectly. Even though the 'International' map sheets would not be used much in India for obvious reasons, the venture would help the Survey of India to remain in touch with modern cartographic developments that were taking place in the world. The authorities thought it would be a good experience to prepare two different series of maps, of the same areas, but in different styles.

The first International Conference on the Map of the World on scale 1:1 M was held in London in 1909. Some specifications about the conventional signs, colour scale for layers and contours intervals, etc., were decided upon regarding the International One-millionth map of the World (generally referred to as IMW). Arrangements for the preparation of sheets, containing territories belonging to several nations, and the nations which had no cartographical establishments, were also discussed.[281]

Government of India received an invitation by telegram, through the Director, Military Operations, War Office, London, from the French Government to attend the second conference, which was to meet in Paris on 10 December 1913, to consider any alterations which might be desirable in the Resolutions which were passed by the similar conference held earlier in London in 1909, for the guidance of those nations who were anxious to assist in the preparation of the International Map of the World (IMW) on the scale of 1:1,000,000 to be represented at the Conference.[282]

Major M.O'C. Tandy, Deputy Superintendent, Survey of India, represented India at the Conference which began on 10 December 1913. On his return, he submitted his report on 19 January 1914.

The chief alterations suggested at the Conference were mostly regarding the conventional signs, contour-scale and colouring, etc. Tentative maps, prepared on the guidelines of 1909 conference, were examined there.

It was also decided to prepare Index charts or accurate drawings illustrating the new symbols to be used on the world maps. An interesting aspect of this Conference was an exhibition held at the Salle d'Honneur, the hall in which all the full meetings of the Conference were held. Copies of all the international 1:1 million maps which were, till then published by various countries, were displayed. These included interesting specimens of maps prepared by England, France, America, Portugal, Japan, Sweden, Spain and Italy. The only map, not belonging to the international series selected for the exhibition, was the Indian 1:1 M, sheet 53 of the post Indian Survey Committee period. This sheet attracted considerable attention. This sheet was the main factor in the decision taken by the International body, after seeing the colours used on the Indian sheet, to employ a bright red colour for the highest layers[283] in the International World map.

The appreciation of the Indian Geodetic work could be assessed from the following 'Note' in the Government of India file appended after the International Geodetic and Geographical Congress held at Rome in May 1922:

> .... From the point of view of geodesy the position of India is unique and she can contribute information which is unobtainable anywhere else, so that the greatest advances of modern geodesy have taken their rise from observations made in this country. The whole science is being pushed forward very rapidly in America, and all civilised Europe, in spite of excessive financial difficulties, is doing a great deal of work in this connection, as may be seen from the proceedings of the International Geodetic and Geographical Congress which took place in Rome last May [1922].[284]

The regular topographical programme of the Survey of India was also disturbed due to the Second World War during 1939-45; otherwise the cycle of completing the topographical mapping on 1-inch scale, in all probability, would have been completed, if not in the stipulated time, much before 1955, when it was actually concluded.

The map sheets covering India, as part of the international efforts to publish 'La Carte Internationale Du Monde', or 'The International Map of the World', on 1:1,000,000 scale, the responsibility of which was accepted by the Survey of India, were completed during 1948-9.[285]

The first ever topographical map of India on 1:1,000,000 scale which was taken up by the Survey of India as a result of the recommendations of the Indian Survey Committee of 1904-5, in spite of tremendous pressures and diversions as a result of other governmental and departmental activities, was completed in 1954-5.[286]

Thus, after when the first cycle was over, the new modern map of India on the metric system began taking shape from October 1956[287] as part of the next revision.

NOTES

1. C.R. Markham, op. cit., 1878, p. 59.
2. Ibid., p. 59.
3. SOI, *D.Dn. Vols.,* op. cit., No. 1820-30/204, p. 88; Letter, dt. 11 Aug. 1824, from the S.G. to Mily. Deptt.
4. *Punna* = Punnae.
5. SOI., *D.Dn. Vols.,* op. cit., No.1820-30/204, p. 88.
6. C.R. Markham, op. cit., 1878, pp. 65-66.
7. Ibid., 1878, p. 66.
8. Ibid., 1878, p. 67.
9. Ibid., 1878, p. 61.
10. SOI, *D.Dn. Vols.,* op. cit., 1820-30/204, p. 89.
11. C.R. Markham, op. cit., 1878, p. 71.
12. R.H. Phillimore, *Historical Records,* op. cit., Vol. IV, p. 2.
13. C.R. Markham, op. cit., 1878, p. 84.
14. Ibid., 1878, p. 86.
15. Public Deptt., Letters from Court, 26 May 1830.
16. SOI, *D.Dn. Vols.,* op. cit., No.1830-40/265, pp. 120-1; Letter No.76, dt. Jan.10, 1831, from S.G. to Mily. Deptt.
17. Ibid., op. cit., No. 1830-40/263, pp. 294-7.
18. Ibid., op. cit., No. 1830-40/265, pp. 261-91; Letter No.171, 12 Oct. 1831, from SG to Mily. Deptt.
19. C.R. Markkham, op. cit., 1878, p. 86.
20. Ibid., 1878, p. 87.
21. Ibid., 1878, pp. 87-8.
22. SOI, *D.Dn. Vols.,* op. cit., No.1830-40/342, Letters, 18 May 1837 and 28 Dec. 1837, from George Everest, SG to the Govt.
23. Ibid., p. 148; Letter, 5 May 1838, from SG to Govt.
24. Ibid., pp. 100-11; Letter 26 Jan. 1838, from SG to Govt.
25. SOI, *D.Dn. Vols.,* op. cit., No.1830-40/341, pp. 202-4, Letter 13 Feb. 1838, from Mily. Deptt. to SG.
26. Ibid., op. cit., No. 1840-50/402, pp. 133-49.
27. Ibid., op. cit., No. 1830-40/343, pp. 364-8;
28. Home Deptt., General, Letters to Court, 30 May 1840.
29. SOI, *D.Dn. Vols.,* op. cit., No 1850-60/401, pp. 116-21; Letter No.592, 26 Jan. 1842, from Mily. Deptt. to SG.
30. Ibid., op. cit., No. 1840-50/402, pp. 256-67; Letter from Mily. Deptt. to Surveyor-General.
31. Ibid., op. cit., No. 1840-50/401, pp. 154-64; Letter, 17 June 1842, from Mily. Deptt. to SG.
32. Ibid., op. cit., No. 1840-50/402, pp. 371-86; Letter, 15 Aug. 1842, from SG to Mily. Deptt.
33. Ibid., op. cit., No. 1840-50/401, pp. 223-8.
34. Ibid., p. 228.
35. Letter from Court to Military Deptt., Government of India, 26 September 1843.
36. Ibid.
37. SOI, *D.Dn. Vols.,* op. cit., No.1840-50/401, pp. 154-64.
38. Ibid., op. cit., No.1840-50/452, pp. 30-6; Letter, 10 July 1843 from SG to Govt.
39. C.R. Markham, op. cit., 1878, p. 102.
40. R.H. Phillimore, *Historical Records,* op. cit., Vol. IV, p. 69.
41. C.R. Markham, op. cit., 1878, p. 103.
42. Ibid., 1878, p. 104.
43. Ibid., 1878, p. 93.
44. Ibid., 1878, p. 103.
45. SOI., *D.Dn. Vols.,* op. cit., No. 1840-50/431, pp, 263-9.
46. Ibid.; Instructions were issued in Oct. 1843.
47. Ibid., p. 104.
48. C.R. Markham, op. cit., 1878, p. 103.
49. Ibid., p. 103.
50. Ibid., p. 104.
51. Ibid., p. 105.
52. Ibid., p. 106.
53. Ibid., p. 106.
54. Ibid., p. 107.
55. Ibid., p. 108.
56. Ibid., p. 123.
57. Ibid., p. 124.
58. Ibid., p. 125.
58. Ibid., p. 125.
60. Revenue & Agriculture Deptt., Surveys, July 1884, Prog. No.13.
61. Ibid., Deptt., Prog. Vol., 1885, Genl. Report, 1884.
62. Ibid., Prog. Vol., 1886, Genl. Report, 1885.
63. Ibid., Surveys, Sept. 1887, Prog. No. 20.
64. Home Deptt., Geographical, Prog. Vol., 1869, 9 Dec. 1868, pp. 166-7.
65. Ibid., Prog.Vol. 1869, Genl. Report, 1868-9.
66. C.R. Markham, op. cit., 1878, p. 127.
67. Ibid., p. 127.
68. Ibid., p. 128.
69. Home Deptt., Geographical, Prog. Vol. 1869, p. 161.
70. C.R. Markham, op. cit., 1878, p. 131.
71. Ibid., p. 131.
72. Home Deptt., Geographical, Prog. Vol. 1869, p. 164.
73. C.R. Markham, op. cit., 1878, p. 131.
74. Home Deptt., Geographical, Prog. Vol. 1869, p. 164.
75. Revenue & Agriculture Deptt., Surveys, June 1883, Prog. No. 8.
76. C.R. Markham, op. cit., 1878, p. 132.
77. Charles E.D. Black, *A Memoir on the Indian Surveys, 1875-1890,* London, 1891, p. 42.
78. Ibid., p. 233.
79. Revenue & Agriculture Deptt., Surveys, May 1882, No.17, Genl. Report, 1880-1.
80. C.R. Markham, op. cit., 1878, p. 74.
81. SOI, *Field-books,* op. cit., No. 60.
82. Ibid., No. 59.
83. Ibid., No. 561.
84. SOI, *D.Dn. Vols.,* op. cit., No.1800-10/68, p. 21.
85. Ibid., No.1820-30/240, p. 104.
86. C.R. Markham, op. cit., 1878, p. 74.
87. S.A.I. Tirmizi, op. cit., 1801-20/4, 16, 17-17A, 57-57A.
88. SOI, *Field-Books,* op. cit., Nos. 210, 210(A), 212, 213, 214.
89. SOI, *D.Dn. Vols.,* op. cit., No.1820-30/246, p. 100.
90. SOI, *Field-Books,* op. cit., Nos.199, 200, 201, 202, 203, 204, 205.
91. C.R. Markham, op. cit., 1878, p. 75.
92. Prasad, S.N.,(Ed.), op. cit., F.133/7, 146/8-10,11.
93. SOI, *Field-Books,* op. cit., Nos.1, 2, 3, 4, 5.
94. SOI, *D.Dn. Vols.,* op. cit., No.1820-30/246, p. 83.
95. Ibid., No.1820-30/246.
96. Ibid., No.1820-30/246; SOI, *Field-Books,* op. cit., No.11; C.R. Markham, op. cit., 1878, p. 76.
97. SOI, *D.Dn. Vols.,* op. cit., No.1820-30/246.
98. Ibid.

99. Ibid.
100. Ibid., p. 69.
101. SOI, *Field-Books*, op. cit., Nos.10,94; SOI, *D.Dn. Vols.*, op. cit., 1830-40/387, p.19; SOI, Prasad, S.N., op. cit., Nos. F.135/12-21, 137/26, 147/26-9, 35-6.
102. SOI, *D.Dn. Vols.*, op. cit., No.1820-30/246, p. 69.
103. Ibid., p. 68.
104. Ibid., p. 72.
105. SOI, *Field-Books*, op. cit., Nos.100-2; SOI, *D.Dn. Vols.*, op. cit., No.1830-40/387, p.161, Letter, 12 Sep. 1838, to S.G.
106. SOI, *D.Dn. Vols.*, op. cit., No.1820-30/246, p. 106.
107. Ibid., p. 115.
108. C.R. Markham, op. cit., 1878, p. 77.
109. This is evident from the data on the maps; Prasad, S.N., op. cit., F.139/5-6, 7-9.
110. SOI, *D.Dn. Vols.*, op. cit., No.1820-30/246, p. 115.
111. Ibid., p. 115.
112. SOI, *D.Dn. Vols.*, op. cit., No.1840-50/452, pp.70-3; SOI, Prasad, S.N., op. cit., F.136/4, 5; 147/2-8.
113. Ibid., No.1840-50/491, pp. 68-9.
114. Ibid., p. 118; Surveyor-General's Report, dt. 6 Oct. 1860; Prasad S.N., op. cit., F. 133/47.
115. Ibid., No.1830-40/328, pp. 27-9; SOI, *D.Dn. Vols.*, op. cit., No.1820-30/246, p. 118.
116. SOI, *Field-Books*, op. cit., No.151; SOI, *D.Dn. Vols.*, op. cit., No.1830-40/327; Prasad, S.N., op. cit., Nos. F.67/32, 68/15, 69/11-22, 70/1-107.
117. R.H. Phillimore, op. cit., Vol. IV, p. 237.
118. C.R. Markham, op. cit., 1878, p. 98.
119. Home Deptt., Revenue, Con., 24 Oct. 1842, No.16.
120. SOI., *D.Dn. Vols.*, op. cit., No.1830-40/316, pp. 69-71; *see* Imperial Gazetteers also.
121. Ibid., No. 1830-40/301, pp. 34-6.
122. Ibid., No.1830-40/321, pp. 285-6; Prasad, S.N., op. cit., F.23/37-8.
123. Ibid., No.1830-40/347, p. 266; Report, dt. 7 Nov. 1838, to S.G.
124. Revenue, Agriculture & Commerce Deptt., Surveys, June 1878, No.6.
125. R.H. Phillimore, op. cit., Vol. IV, pp. 233-4.
126. Military Deptt., Con. No.139, 3 Sep. 1832, Letter from Court, 26 May 1830.
127. R.H. Phillimore, op. cit., Vol. IV, p. 303; Military Deptt., Letter from Court, 20 July 1831.
128. SOI, *D.Dn. Vols.*, op. cit., No.1830-40/267, pp. 142-43; SOI, Ibid., No. 1830-40/286, p. 286; R.H. Phillimore, op. cit., Vol. IV, p. 261.
129. Ibid., op. cit., No.1830-40/265, p. 118.
130. Ibid., No.1840-50/402, pp. 197-207.
131. Ibid., No.1830-40/343, pp. 369-76.
132. C.R. Markham, op. cit., 1878, p. 181.
133. SOI, *D.Dn. Vols.*, op. cit., No.1840-50/459, pp. 78, 231-35.
134. Prasad, S.N., op. cit., F.5/1-2; India Office, A Catalogue, op. cit., No.1830-40/265, pp. 227-9.
135. SOI, *D.Dn. Vols.*, op. cit., No.1830-40/265, pp. 227-9.
136. R.H. Phillimore, op. cit., Vol. IV., p. 297.
137. Home Deptt., Revenue, Sept. 1839, No. 260.
138. SOI, *D.Dn. Vols.*, op. cit., No.1830-40/343, pp. 289-95.
139. Ibid., pp. 289-95.
140. Ibid., No.1840-50/408, pp. 36-9.
141. Ibid., No.1840-50'410, Letter, 13 Aug. 1840 from DSG to SG.
142. Ibid., No. 1830-40/283, Letter 3 Oct. 1832, from SG to Govt.
143. Ibid., No.1840-50/410, pp. 70-73; Military Deptt., Letter from Court, 4 Nov. 1841.
144. W.M. Coldstream, *Records of the Survey of India, Vol. XII, Notes on Surveys of India Maps* (Calcutta, 1919), p. 1.
145. Susan Gole, *Early Printed Maps*, op. cit., Preface.
146. C.R. Markham, op. cit., 1878, p. 177.
147. Home Deptt., Geographical, 25 Feb. 1871, 'A', No.9; Letter No.196, 3 Feb. 1871, from SG to Govt.
148. Ibid., 25 Feb. 1871, Prog. No.10.
149. Revenue & Agriculture Deptt., Land Surveys, Aug. 1901, Prog., No.14, p. 231.
150. Ibid., Land Surveys, June 1878, Prog. No. 6.
151. Ibid: Letter No. 243, 5 June 1878.
152. W.M. Coldstream, op. cit., p. 13.
153. Revenue & Agriculture Deptt., Land Surveys, Aug. 1901, No.14, pp. 234-5.
154. Ibid., p. 235.
155. Revenue, Agriculture & Commerce Deptt., Surveys, Aug. 1874, Prog. No. 8.
156. Ibid., Surveys, Dec., 1874, Prog, No.10.
157. Survey of India, Genl. Admn. Report, 1902-3.
158. Revenue & Agriculture Deptt., Land Surveys, Aug. 1901, Prog. No.14, pp. 235-6.
159. Imperial Gazetteer of India, 1908, Vol. X, p. 98.
160. Ibid., pp. 98-9.
161. Ibid., pp. 387.
162. Ibid., p. 387.
163. Ibid., pp. 387-8.
164. Ibid., p. 388.
165. Ibid., 1908, Vol. XXI, p. 158.
166. According to C.R. Markham, it was introduced in 1836.
167. Imperial Gazetteer of India, 1908, Vol. VIII, p. 381.
168. C.R. Markham, op. cit., 1878, p. 195.
169. Home Deptt., Geographical, 22 April, 1871, Prog. No. 10; Letter No. 324, 22 Feb. 1871, from SG to Govt.
170. 1866, according to Imperial Gazetteer of India, 1908, Vol. VIII, pp. 381-2.
171. Revenue & Agriculture Deptt., Surveys, Nov. 1882, Prog. No. 12.
172. Ibid., Land Surveys, Aug. 1901, Prog. No. 14.
173. Ibid., Surveys, Nov. 1882, Prog. No.12.
174. Imperial Gazetteer of India, Vol. VIII, 1908, p. 382—'By 1904, nearly the whole of the Presidency had been mapped'.
175. C.R. Markham, op. cit., 1878, p. 180.
176. Ibid., pp. 188-9.
177. Home Deptt., Geographical, 22 April 1871, Prog. No. 10; Letter No. 324, 22 Feb. 1871, from Surveyor-General to the Home Deptt.
178. C.R. Markham, op. cit., 1878, p. 189.
179. Ibid., p. 189.
180. Home Deptt., Geographical, 30 Oct. 1869, Prog. No. 3.
181. Imperial Gazetteer of India, Vol. XVI, 1908, p. 348.
182. Revenue & Agriculture Deptt., Land Surveys, Aug. 1901, Prog. No. 14.
183. Imperial Gazetteer of India, Vol. XVIII, 1908, p. 250.

184. Revenue & Agriculture Deptt., Land Surveys, Aug. 1901, Prog. No. 14.
185. C.R. Markham, op. cit., 1878, p. 175.
186. Imperial Gazetteer of India, Vol. XIII, 1908, p. 299.
187. Revenue & Agriculture Deptt., Land Surveys, A, Aug. 1901, Prog. No. 14.
188. Ibid., Prog. No. 14.
189. Imperial Gazetteer of India, Vol. XIII, 1908, p. 299.
190. Revenue & Agriculture Deptt., Land Surveys, A, Aug. 1901, Prog. No. 14.
191. C.R. Markham, op. cit., 1878, p. 178.
192. *India Office, A Catalogue.*, op. cit., 1878, pp. 589-91.
193. Susan Gole, *Early Printed Maps*, op. cit., Sl. No. 56, Map. 56.
194. Revenue, Agriculture & Commerce Deptt., Surveys, Vol. March 1877, Appendix B; Gen. Report of the SOI., 1875-76, 4 Dec. 1876.
195. W.M. Coldstream, op. cit., p. 11.
196. Home Deptt., Geographical, A, 30 Oct. 1869, Prog. No.3.
197. Revenue & Agriculture Deptt., Land Surveys, A, 31 Oct. 1904, Prog. No.19.
198. Ibid., Land Surveys, A, Aug. 1901, Prog. No.14.
199. Imperial Gazetteer of India, Vol. VII, 1908, pp. 339-411.
200. Compiled from the SOI, Rev. Traverse Records in the NAI, and the Imperial Gazetteer of India, Vol. XXIV, 1908, p. 256.
201. Revenue & Agriculture Deptt., Land Surveys, A, Aug. 1901, Prog. No. 14.
202. Data has been compiled from the SOI, Revenue Traverse Records in the NAI; Revenue & Agriculture Deptt., Land Revenues, 'A', File No.164/Sl. No. 1, 7 Feb. 1900.
203. Ibid., Surveys, A, Sept. 1887, Prog. No. 20; Gen. Report on the Operations of the Survey of India, 1885-6.
204. Ibid., Land Surveys, A, Aug. 1901, Prog. No.14.
205. Ibid., Surveys, A, Sept. 1887, Prog. No. 20; Gen. Report, on the Operations of the Survey of India, 1885-6.
206. Ibid., Land Surveys, A, Aug. 1901, Prog. No.14.
207. Revenue, Agriculture & Commerce Deptt., Surveys, March 1878, Prog. No. 3; Letter No. 503, 3 March 1878, from the Surveyor-General of India to the Govt.
208. Ibid., Surveys, March 1878, Prog. No. 4; 21 Letter March 1878, No. 172, from the Govt. to the Surveyor-General of India.
209. Ibid., Surveys, A., Jan. 1879, Prog. No.1.
210. Ibid., Prog. No.1.
211. C.E.D. Black, op. cit., p. 102.
212. Revenue Agriculture & Commerce Deptt., Surveys Jan. 1879, Prog. No.1.
213. C.E.D. Black, op. cit., p. 102.
214. Revenue, Agriculture & Commerce Deptt., Surveys, A, Jan. 1879, Prog. No. 1; Letter No. 295F., 22 Aug. 1878, from the Surveyor-General of India, to Govt.
215. C.E.D. Black, op. cit., p. 103; Revenue, Agriculture & Commerce Deptt., A, Jan. 1879, Prog. No.1.
216. Revenue & Agriculture Deptt., Surveys, A, 4 Sept. 1882, Prog. No.14.
217. Ibid., Surveys, A, Sept. 1882, Prog. No.12.
218. Ibid., Surveys, A, Sept. 1882, Prog. No. 14; Circular No. 45-S.
219. Ibid., Surveys, A, Nov. 1882, Prog. No.12.
220. Ibid., Surveys, A, Aug. 1883, Prog. No.13.
221. R.H. Phillimore, op. cit., Vol. II, p. 10.
222. Ibid., Vol. II, p. 12.
223. C.R. Low, op. cit., Vol. II, p. 396.
224. Ibid., Vol. II, p. 400.
225. Ibid., p. 400.
226. Ibid., p. 402.
227. Ibid., p. 394.
228. Ibid., p. 394
229. Ibid., pp. 394-8.
230. C.E.D. Black, op. cit., p. 1.
231. Revenue, Agriculture & Commerce Deptt., Surveys, A, March 1877, Appendix A.
232. Ibid., Appendix A.
233. Revenue, Agriculture & Commerce Deptt., Surveys, Despatch to SOS, No. 17, 1 May 1874.
234. Ibid., Surveys, A, March 1876, Prog. No. 6; Despatch No.12 (Geographical), India Office, London, 16 July 1874.
235. Ibid., Prog. No. 6.
236. Ibid., Prog. No. 6.
237. Ibid., No. 6; C.E.D. Black, op. cit., p. 16.
238. C.E.D. Black, op. cit., p. 4.
239. Revenue, Agriculture Deptt., Surveys, A, Jan. 1883, Prog. No. 14.
240. Ibid. Prog, No. 14.
241. Revenue & Agriculture Deptt., Land Surveys, A, Aug. 1901, Prog. No. 14; Letter No. 139, 5 June 1901, from the Surveyor-General of India to the Govt. of India.
242. Ibid., Land Surveys, A, Aug. 1901, Prog. No.15; Letter No. 1803, 7 Aug. 1901, from the Govt. of India to the Surveyor-General of India.
243. Ibid., Land Surveys, A, July 1904, Prog. No.14; Letter No. 963, 12 July 1904, from the Govt. of India to the Surveyor-General of India.
244. Ibid., Land Surveys, A, Sept. 1904, Prog. No. 1; Circular No. 2-70-9, 7 July 1904.
245. Ibid., Land Surveys, A, Sept. 1904, Prog. No. 3; Letter to SOS, 1 Sept. 1904.
246. Ibid., Land Surveys, A, 31 Oct. 1904, Prog. No. 19.
247. Ibid., Land Surveys, A, Sept. 1904, Prog. No. 3.
248. Imperial Gazetteer of India, 1908, Vol. IV, pp. 506-7.
249. Revenue & Agriculture Deptt., Land Surveys, Feb. 1906, Prog. No.10; Letter No. 4621, 23 Dec. 1905, from Surveyor-General of India to the Govt. of India; Revenue & Agriculture Deptt., Land Surveys, Sept. 1909, Prog. No. 17; Letter No.152 (Revenue), 25 Dec. 1908, from SOS to the Govt. of India.
250. Revenue & Agriculture Deptt., Land Surveys, Oct. 1908, Prog. No. 15; Letter No. 749, 31 Aug. 1908, from the Surveyor-General of India to the Govt. of India.
251. W.M. Coldstream, op. cit., pp. 29-31.
252. Revenue & Agriculture Deptt., Land Surveys, February 1908, Prog. No. 10; Letter No. 4621, 23 December 1905, from the Surveyor-General of India to the Govt. of India.
253. Letter No. 262, 27 August 1908, from the Finance Deptt. to the SOS.
254. Revenue & Agriculture Deptt., Land Surveys, February 1906, Prog. No. 10; Letter No. 4621, 23 December 1905, from the Surveyor-General of India to the Govt. of India.
255. Revenue & Agriculture Deptt., Land Surveys, February

1912, Prog. No. 1; Letter No. 55T, 8 July 1911, from the Offg. Surveyor-General of India to the Govt. of India.
256. Ibid., Land Surveys, February 1912, Prog. No. 3; Circular No. 3-110.3, 10 February 1912, from the Govt. of India to all state Govts. and Administrators.
257. Ibid., Land Surveys, 17 June 1915, Prog. No. 6.
258. Rev. Despatch No. 130, 9 August 1907 to the SOS.
259. Revenue & Agriculture Deptt., Land Surveys, A, March 1913, Prog. Nos. 29-56, F. No. 46 of 1912.
260. Ibid., Land Surveys, A, March 1915, Prog. No. 2, F.100 of 1914.
261. Ibid., Land Surveys, A, March 1915, Prog. No. 3.
262. Ibid., Land Surveys, A, May 1915, Prog. No. 4; Letter No. 245-T, 6 May 1915, from the Surveyor-General of India, to the Govt. of India.
263. Ibid., Land Surveys, A, March 1913, Prog. Nos. 29-56; Despatch No. 4-Rev., 23 January 1913, to the SOS.
264. Ibid., Land Surveys, A, June 1915, Prog. No. 5.
265. Ibid., Land Surveys, A, September 1908, Prog. No. 3.
266. Ibid., Land Surveys, A, January 1913, Prog. Nos.1-3 (F. No. 145, 1912).
267. Ibid., Land Surveys, A, September 1913, Prog. Nos. 10-12 (F. No. 91).
268. Ibid., Land Surveys, A, May 1917, Prog. Nos.1-2 (F. No. 25).
269. Ibid., Land Surveys, A, September 1908, Prog. Nos. 19-26 (F. No.57); ibid., Land Surveys, A, October 1915, Prog. No. 1-9, (F. No. 37); ibid., Land Surveys, A, November 1916, Prog. Nos. 3-10, (F. No. 8); ibid., Land Surveys, A, November 1917, Prog. Nos. 3-7 (F. No. 51).
270. Ibid., Land Surveys, A, June, 1914, Prog. No. 15-16, (F. No. 65); ibid., Land Surveys, A, January 1917, Prog. Nos. 8-29, (F. No. 65 of 1914).
271. Ibid., Land Surveys, A, May 1913, Prog. Nos. 7-21.
272. Ibid., Prog. No. 7-21.
273. Ibid., Land Surveys, A, May 1911, Prog. No. 3.
274. Ibid., Land Surveys, A, July 1911, Prog. No. 23; Letter No. 81/949 (M.O.4.) White Hall, S.W., London, 22 February 1911, from Director Military Operations to Under Secretary of State for India.
275. Ibid., Land Surveys, A, February 1906, Prog. Nos. 19,11; Letter No. 1621, 23 December 1905, from the Surveyor-General of India to the Govt. of India.
276. Letter No. 82-149.2, 19 Jan. 1906, from Govt. of India to the Surveyor-General of India.
277. Letter No. R. and S, 4887, India Office, London, 22 March 1911, from SOS, to the Govt. of India.
278. Letter No. 81/849 (M.O.4.), 2 April 1911, from the Director Military Operations, War Office, London, to the Govt. of India.
279. Letter No. 21, 20 July 1911 from Revenue & Agriculture Deptt., Land Surveys, Govt. of India to the SOS.
280. SOI, Chart, *Table of Lettering & Symbols for International Map of the World, 1: One Million*, 1952 Edn.
281. Revenue & Agriculture Deptt., Land Surveys, A, February 1914, Prog. No. 3.
282. Ibid., Land Surveys, A, October 1913, Prog. No.15; Telegram from the Viceroy, Simla to SOS, London, 25 October 1913.
283. Major Tandy's report on the International Conference, on the Map of the World, on scale 1/1,000,000, held in Paris during December 1913; Revenue & Agriculture Deptt., Land Surveys, A, February 1914, Prog. No. 3.
284. Revenue & Agriculture Deptt., General, Field Papers, August 1923, Prog. No. 54 (File No. 87 of 1922).
285. Survey of India, General Report, 1948-9.
286. Survey of India, General Report, 1954-5.
287. Ibid., 1956-7.

7. Indigenous Maps

While going through the various stages in the evolution of the modern 1:1 M Topographical map of India, it is but natural to ask whether India made any contribution (a) to the methodology of map making, (b) to the correct delineation of the shape of India, particularly the coastline and the topographical features of the country, and (c) whether any ideas relating to map-making developed here independently or otherwise of the progress made in this field.

Unfortunately, no contemporary literature on the subject, either by way of written manuscripts or inscriptions has come to light so far which would satisfactorily answer these queries. Early Indian writers, devoted much of their energies to the exposition of Indian religious and philosophical thought and astronomical mathematics. The history of the evolution of the Indian literature in Sanskrit, Pali and in other ancient languages, including Tamil and other Southern languages goes to prove this. Later in the period, mention is made in the Buddhist literature of the Indian voyages to various littoral countries of the Indian Ocean, especially to the South and South-East Asia. But this literature also does not help in arriving at any conclusion in regard to the problems posed above.

Strange though it may seem, in ancient Indian languages, there is no contemporary equivalent word for the English word 'map'.[1] Some writers have explained this by saying that, in the early period of India's history, the making of maps was considered a part of the art of painting. Others, however, say that map-making was not known to early Indians, and as such it was either covered by the word 'Chitra', or it did not exist at all.[2] All this points to the conclusion that there was no worthwhile Indian contribution to the methodology of map making.

However, we have also seen that in ancient times, the people of India had mastered the knowledge of the air and sea currents and were far superior to the rest of the known world in the naval techniques. Sufficient proof is available that Indians have contributed much in the development of the navigational sciences, nautical charts, ship-building, etc. Evidence of their contribution is still preserved not only in the story-books of Buddhists and other south Indian literature like *Kathasaritsagara, Jatakamaka,* etc., but is also preserved in folk memory and cultural traditions in poetry, songs, etc., of the fisher-folks living all along the Indian coastal belt more especially on the entire western, southern and south-eastern part of India.[3]

In addition, a number of hints have also been thrown up by many of the early Persian, Arab, European and Chinese travellers, writers of travel accounts and even known cartographers about the existence of Indian maps and charts, though not of the type which these peoples were used to, detailed in the early part of the narrative.

Taking a clue from them, during the last fifty years some scholars, both Indian and foreign, have brought to the notice of the cartographers in particular and the enlightened public in general, their discoveries of many indigenous 'map-illustrations'. This has created a hope of ultimately proving that India too had, like the rest of the early contemporary civilizations, a continuous tradition of map-making.

In an article[4] on 'Maps and Charts' in the *Bharata Itihasa Samsodhaka Mandala* (1948), Poona, Prof. D.V. Kale talks of about two dozen maps held by the Mandala. Though not dated, the maps are believed to be of eighteenth century. Some of these maps are of the period of the Peshwas and the others belong to early British rule just after the defeat of the Peshwas. Prof. Kale has classified his find into three categories.

The first category consists of military maps and charts depicting the formations of the army at war, or the *Vyuhas,* so essential in the ancient warfare when the army consisted of 'foot soldiers, horse soldiers, the elephant soldiers and the charioteers'. These maps might have been in use in the early Maratha uprisings, but would not have been useful later when artillery was introduced in the battles. The *Vyuha* charts are eight in number and they show six different varieties of *Vyuhas,* i.e. the *Mala Vyuha* (or garland), the *Shakata Vyuha* (or cart), *Danda Vyuha* (or bar), *Makara Vyuha* (or crocodile), *Garuda Vyuha* (or eagle) and *Suchi Vyuha* (or the needle). For explaining these *Vyuhas,* every chart carries a Sanskrit *sloka* in one of the top corners from some ancient texts.

Prof. Kale has classified six maps out of his find as 'ground plans or charts of forts'. The forts described are Chandan, Sajjangad, Nandgiri, Adgad, and Vandan. The sixth could not be identified. The village maps and other maps covering wider areas, encompassing a number of forts and villages, and some of them even covering the whole of the southern India, south of the latitude of Poona, are all in *Modi* script and bear no scale.

These 'linear drawings' seem to be a temporary device to 'note cotemporary administrative details'. For example, along with the symbol of each watch-post shown on the map, the number of watchmen keeping watch and also the name of the head watchman are written in words. These maps were useful aids to the day to day administration of the forts. Prof. Kale's evaluation of the maps is worth quoting:

... These are but linear pictures with the free use of lettering, not only to indicate proper names, but even to add the necessary explanations regarding such inferrable data as the relative distances between places and places. The directions are broadly true, and usually noted in as many sentences, main gates and bastions roughly in place, trees, tanks, temples, houses of officers and office halls and standard posts are tolerably in position.... All the distances are maked in the *gaja* (yard) without any fraction anywhere....[5]

The remaining maps are 'village plans or country maps'. These are either mere linear pictures or coloured representations. In these maps, every thing is explained in words, nothing is left to imagination and conventional signs. The whole performance is amateurish. According to Prof. Kale: 'These have neither the plainness nor the bold finish of the simple diagrams of the earlier period and are the bastard progeny of the mixture of the unappreciated eastern simplicity.[6]

All the maps are undated. Wherever some place names or other objects are explained, these are written in *Modi* script, which shows that these belong to the Peshwa period; no scale is used as the scale, according to the author, is a considerably later development. Also, nowhere on any map is the name of the 'artist' or the 'cartographer' mentioned.[7]

In 1951 D.C. Sircar, in an article[8] described many sources and tried to prove that in ancient India, there were many cosmological maps of India, based on *Puranic* literature dealing with the science of cosmology in India.

About the geographical and topographical maps, Sircar is not forthright in his assersion about their existence in early India. But he categorically stated that there is no special word in Sanskrit for a map. He was equally positive and tried to prove that the word *Chitra* or *alekhya* covered the 'map' as well, since in ancient India a 'map' or 'chart' was regarded as a *chitra* or *alekhya*. He further asserted that since the Arabic word '*naqshah*' has the same meaning and connotation, why not the word 'chitra'?

To prove the existence of geographical or topographical maps and their being called paintings, Sircar quotes Bhavabhuti of the eighth century A.D. In 'The inspection of the painting', in Act I of the *Uttrasamacharita*.[9] He describes some paintings drawn along a long walk, all of them though individual in character, were exhibited in a sequence. The paintings depicted the 'experiences' of king Rama of Ayodhya in Dandakarnya, Kishkindhya, Lanka, etc. Some of the paintings, of course, showed the mountainous landscape. Sircar considered these map-like paintings of the eighth century depicted in the Sanskrit drama to be 'maps'.

Sircar also mentioned, though indirectly, the delineation of India's coastline and the improper delineation in Ptolemy's *Geographia* and the reasons thereof. He has also proved the existence of the Indian sea charts having close network of lines intersecting each other at right angles. The horizontal lines, according to him, are parallels depending upon the altitude of the Pole Star, the calves of the Little Bear and Barrow of the Great Bear visible above the horizon.

In 1952, there appeared on the Indian scene, an analytical article by the late Colonel R.H. Phillimore of the Survey of India, a trained and practical cartographer. He analysed his find of indigenous maps in a truly professional manner. The article[10] is entitled 'Three Indian Maps'. The first two maps are (i) Punjab (Map of the Punjab, Sind, and portion of Kabul and Baluchistan, in English and Persian). No scale. Not dated. Mss and 4 Bromide prints, 2 kodalines,[11] and (ii) 'Sketch of Goa by a native of the country. No scale. Not dated. Mss with 2 Bromide Prints'.[12] These are listed in the catalogue compiled by the National Archives of India. These, along with another 'A Bengali sketch of part of the Boundary of Pargana Habraghat, Lower Assam'[13] are the subject matter of Phillimore's article.

The first map, according to Phillimore, is a copy made from the original in about 1780. The original, as per his calculations, was probably made between 1650 and 1730. The map covers the area of part of Afghanistan, Punjab, Rajputana, Sind, the territory between Delhi and Kabul, and areas from Kashmir to the Arabian Sea. The names of places are in Persian and the English rendering have been added. Apart from other details, the map shows the main rivers without any attempt at exactness. The 'wildness in geographical positions' of places and general topography is discernible all over. Notes on the nature of the topography, that would be of the greatest value, especially to the military commanders in the event of a war, are written at many places. It is generally a road map intended for military purposes. About the question as to who made this map, an Indian native or a Persian at Akbar's court or someone else, Phillimore does not even make a wild guess. The general appearance of the map, however, indicates that the map was definitely not prepared by a professional map-maker, and in those days the only professionals in the field were Europeans.

Above all, the map lacks any clear concept of scale or direction. The map, as Phillimore stated 'is generally sadly out of true geographical layout and this has led to confusion in all cross routes'. Had there been any tradition of topographical or terrestrial map-making in

India's long period of civilization since antiquity, the state of Indian indigenous maps would not have been so poor at the court of one of the greatest rulers in India.

While critically examining the above map, Phillimore incidentally referred to another 'Persian Ms map of the Punjab' used by James Rennell which 'was drawn by a native, and preserved in the archives of the Government in Hindustan'.[14] Phillimore has categorically stated that the map of Punjab examined by him and the one used by J. Rennell, though bearing the same title, were two different maps.[15]

The other map under the scrutiny of R.H. Phillimore—Sketch of part of the Boundary of *Pargana* Habraghat (Goalpara District), Lower Assam, is a very rough sketch. It looks like a scribbling done on a page of an official file, dated 1 April 1849, by the Deputy Collector, an Anglo-Indian, educated in English tradition and posted at *Dookahpara*. The sketch was meant to indicate to the professional surveyor the general position of the lands of which the boundaries were to be surveyed. The approximate positions of the plots of land were roughed in by the Deputy Collector and lettering in Bengali script was definitely done by a Bengali clerk in the office. It was by no means a map of much value in the 'chain' of the evolution of the Indian mapping in the mid-nineteenth century period.

The third map that was evaluated by R.H. Phillimore was a Maratha map of North Kanara region. This map is now in the National archives of India. Phillimore has stated that 'it is very rough and untidy, with names and descriptive notes in the Maratha language in the *Modi* script....' The official label describes it as 'Sketch of Goa by a Native of the country'. But actually, though the south boundary of Goa was intended, the map covers the north coast of North Kanara southwards up to the Aghnashini river. The map stretches about 35 miles along the coast and about 12 miles in-land. The main coastline, the rivers, roads, limits of villages, jungle, hill, paddy fields and gardens are roughly sketched.[16]

One thing that struck Phillimore most was that the *Modi* names are well and clearly written in strong black ink whereas the outline of the map and the English renderings were faded and badly written and difficult to read.

To explain his surmise that the map was probably drawn towards the end of the eighteenth century and might be a copy of the original, he delved into the past history of the region. He noted that North Kanara was conquered by the Marathas before the middle of the eighteenth century. Shortly after, Tipu Sultan occupied it. It came into the hands of the English after Tipu's defeat and death in March 1799. South Kanara was surveyed by Colin Mackenzie between 1801 and 1806, but North Kanara was not surveyed till some eight years later. This rough map, under study, might have been prepared for the British Collector of 1799, whom it would have provided the most valuable information.

Next in the chain of further discovery of some more terrestrial maps, is C.D. Deshpande, who came out with his descriptive article in 1953. He described some more Maratha maps,[17] other than those which Prof. Kale had dealt with earlier. These maps are also written in *Modi* script, but in some cases, traces of Devnagri additions have also been noted.

Some maps, generally fort plans, are very well explained and described by the map-makers (or might be, a single map-maker was involved as the style of writing and map-decorations seem to be from the same hand) with descriptive notes which are profusely written on the maps themselves; the distances are given in Cos. On the other hand, on some other maps the information given is very vague. Deshpande describes the following fort plans, Vijaydurg, Budargarh, Janjira, Badami and Seringapatam. On some regional maps, cartographic features are given in some details. This is also true of a map wherein no title is given. This covers the area between West sea coast, Krishna and Tungabhadra rivers. In all, he details the information discernible from other maps of Gujarat-Kathiawar region, the Bhima-Malaprabha river valleys, and maps of the areas of the Nizam's dominions during the end of the eighteenth century (as this is the probable date of the map).

Another such Maratha map, depicting larger area, is at present lodged in the Bhiwalkar collection, Maratha History Museum, Deccan College, Pune.[18] This map is also untitled and is not drawn to scale. The map is north-oriented and covers the area between Bhimashankar in the bottom left corner on the Bhima river, Shahgarh on the Godavari river in the bottom right corner, Satpura mountain ranges on the top left corner, north of Tapti river, and Mahalgudad east of Burhanpur and Jainabad, also on the Tapti river, in the top right corner.

Almost all the major towns and places in the region are shown on the map, like Poona, Malegaon, Songiri, Jalgaon, Aurangabad, Malwapur, Jafferabad, Ajanta, Akola, Burhanpur, etc., and the rivers like Tapti, Goadavari and Bhima along with Satpura mountains are depicted. At two places, one just south of Jafferabad, and the other at south-east of Malwapur, the word 'battle' is written, indicating thereby that 'battles' took place there. During 1803-6 Colonel Sir Arther Wellesley, brother of the Governor-General Lord Richard Colley Wellesley, along with Stevenson, engaged the Marathas in the region. This map was probably made during or after the Maratha wars.

As the scale and the compass were not used to prepare the map and with un-measured distances, the relative positions of the places are somewhat different from their geographical positions and directions.

In 1980 Prof. Irfan Habib in his article on 'Cartography

in Mughal India',[19] brought out the salient features of the *Shahid-i-Sadiq* by Sadiq Isfahani. This encyclopaedic work, a world atlas in Persian, was completed by him in 1647 at Jaunpur, in modern Uttar Pradesh. The article has not been illustrated. This cartographic marvel on the Indian scene was unique in that its maps were not drawn in the known and, by then, more familiar style of terrestrial representations or projections. It was based on, what in cartographic jargon is called 'Cylinderical system of projection'. The only other cartographer known at that time for such maps was Hamdullah Mustaufi[20] of the early fourteenth century who had drawn maps of Iran. The main difference between this and other known projection systems is that in the cylinderical projection the map is based on coordinates only, i.e. on the latitudes and longitudes of the places, and route distances and true directions are not taken into account. Even in the coordinates, Sadiq did not make use of minutes but considered only degrees while drawing his maps. His maps, therefore, do not show routes. This is perhaps the only such work on India done in India. These maps are certainly not without foreign influence. However, this atlas, though difficult to read, is quite unique in itself.

In the same article, Prof. Habib has also touched upon another unique work, the *Chahar Gulshan,* or 'Four Gardens'. This is a short history of India in the form of a number of itineraries wherein topographical information was taken from the *Ain-i-Akbari*. This unique work was written in 1760 by Rai Chaturman Saksena.[21] According to Prof. Habib it did not contain mere lists of names of places serving as stages, with distances between them but had 'cartographic elements too, with courses of rivers drawn to run across the routes and ... mountains shown by drawing short vertical lines along a few parallel horizontal lines', in a most elementary fashion.

More recently, Susan Gole, a scholar, a prolific writer and an enthusiastic collector of old maps of India in her own right, has collected over hundred 'indigenous maps' of various types from widely scattered places in India, covering the period from the late medieval to the early colonial days. This discovery of so many 'map illustrations' has further strengthened the hope that India too must have had a tradition in map-making.

In her book *Indian Maps and Plans* which is the fifth in the series, Susan Gole has examined and commented upon the works and views of almost all the known writers on the subject till about 1989, when the book came out. She had also included most of their maps in her volume.

A perusal of the book shows that in the early period of the cartographic history the map-illustrations contained just graphic descriptions of the topographic features as well as mental constructs of what the map-makers had observed of the earth. The contents of Susan Gole's compilation could well be understood or evaluated by classifying 'the map illustrations' into broad groups like—religious, topographical, military maps, sea-charts, towns and constructions maps and plans. Also included are a few early Indian globes, mostly cosmological in character.

The religious 'maps' consist of plans of temples like those at Jagannath Puri, Kashi, Dwarka, etc. Yet another religious 'map illustrations' are *Vijnatipatras* or *Tirthpatras* of Rajasthan and Kashmir. These are like invitation cards of modern days, but in the form of painted scrolls showing landmark buildings, all looking alike, along the route leading to a particular place. These are not dated, but are believed to be of late eighteenth and early nineteenth centuries. These are highly imaginative, having no regard to scale and direction. These have no likeness at all to actual shapes of the buildings or temples they suppose to represent. The *Vijnatipatras* seem to be the works of artists and not of map-makers.

The military maps are generally of battle plans or siege of some forts and layout plans of the forts in Rajasthan and Maharashtra. Those from Rajasthan have more decorative elements than the technical requirements of the theme.

Fort plans generally show the walls in elevation, devoid of perspective techniques. Inside the fort walls, the approximate areas covered by the less important buildings are shown in rectangles; the more important buildings are marked by the imaginary plinth shapes, as conceived by the artist, without caring for their scale. But in the maps covering the larger areas of the town, major streets are marked with built-in areas of residential localities. The streets cross each other at right angles.

The 'construction' maps and plans are rough sketches of town and fort layout plans, mostly from Rajasthan, with characteristics similar to the Rajasthani style of imaginative and highly decorative paintings.

About maps in *Chahar Gulshan,* Susan Gole has not added much topographical information other than what Prof. Habib has given in his article, except to say that this compilation was also known as '*Akhbaru-l-Nawadir* or the 'Accounts of Rare Things'.[22] She has also included the maps in her book. Likewise, she has also incorporated many of the *Pargana* maps[23] of Kashmir as given in the history of Kashmir, *Tarikh-i-Kala,* of the early nineteenth century, authored by Miham Singh. The main emphasis, as is apparent from the maps, is on the rivers, and mountain ranges. Very few large buildings are shown in elevation. The artist who drew the 'map paintings' seems to have seen foreign maps, particularly maps with Persian influence, and had tried to imitate them in his own way, without a rudimentary knowledge of the application of scale, direction and drawing of other topographical details. Uniformly depicted, in wide serpentine forms, are the rivers, almost covering the entire map, with their imaginary courses. Mountain ranges, shown in elevation, are amateurish in style and drawing. The remaining blank spaces in the map, are filled by place names, in Persian script with circles encompassing each of them. Scanty vegetation is seen on some maps. Surprisingly, no roads are shown in these *Pargana* maps, and no other topographical features could be discerned.

On the other hand, the individual terrestrial maps of Kashmir, Gujarat and Rajasthan and of the southern and western India included in Gole's book, are remarkably well drawn and unmistakably show foreign influence which is perceptible all over. These maps generally cover large areas, encompassing some villages and forts or even large parts of the states like Gujarat, Rajasthan, Central India, Maharashtra, etc. They seem to be more utilitarian in nature, though the maps of Kashmir appear otherwise. Almost all such terrestrial maps are probably of the nineteenth century.

Going through this valuable book, one cannot miss the point that almost all the maps discussed, belong to the western and northern India, west of Bihar, with only a few that relate to southern and eastern India. A closer scrutiny of the maps also reveals that there is no consistency or chain of thought in the evolution of the topographical map of India.

These maps, collected from different parts of India, are isolated cases and seem to display more than anything else, religious veneration and sentiments for the overlords who ordered the maps to be prepared. Most of the maps do not give proof of dissemination of any historical cartographic ideas. On the other hand, even on their face value, without knowing the period to which they belong, they give an idea of the early Indian style of two-dimensional imaginative, and artistic representation of landscapes by artists without caring for or having proper sense of direction or of proportional distances between places and shapes of topographical features. However, in some cases, mostly in the Maratha cartographical sketches, some consistency of topographical representational methodology could be discerned.

By the end of the eighteenth century many Indian rulers and military officers, especially in the south, including influential people had come in contact with the Europeans. They, no doubt, had seen the latter using maps on many occassions and wondered, in the beginning, as to why maps were being consulted by them for both war and peacetime activities. The English asked permission for survey and demarcation of boundaries of the territories of their allies from many of the native rulers, like the Raja of Carnatic. This was very common at that time in the South. Many inquisitive eyes must have wondered as to how these maps were prepared. By merely looking at a map closely, it would not have been difficult for an intelligent man to grasp the rudiments of a map. But lack of proper training in compass application, in the use of scales and distance measurements, etc., probably resulted in the maps, like the Maratha maps of larger territories in *Modi* script, discussed earlier. One cannot say, therefore, with certainty that these maps were uninfluenced by the European map-making traditions. Clearly, by the early nineteenth century, geographical curiosity was definitely gradually dawning upon the responsible and learned sections of the society in the Maratha region.

The author has in his possession an old *Marathee Atlas*[24] published in 1836 at Bombay. It contains nine maps. The maps are political in nature, drawn and published by Dadoba Pandurung and Nana Narayun, and lithographed at Bombay by Gunput Crustnajee in 1835-6. The atlas is 'dedicated' to 'Balcrushna Gungadhur and Vinayeck Gungadhur, sons of the late Gungadhur Shastree, Putvurdhun, the distinguished friends of Native Improvement, the following humble attempt, to diffuse Geographical Knowledge among the natives of India, is, with the most affectivenate [This is perhaps a printing mistake, could be 'affectionate'] regard, inscribed, by their friends. The publishers. Bombay 1st June 1836.'

The maps are hand-coloured, lithographed maps of the world, Asia, Europe, Africa, North America, South America, Hindustan (India), Maharashtra Region and of Great Britain. Some of the maps are drawn to scale, with the prime meridian passing through Greenwich. There are six pages of text which include elementary definitions of geographical terms and 'Preface' in a page and a list of 'Abbreviations' used on the maps. It also contains two pages of lists of countries in Asia and Europe, giving the names of important cities in each country with its area and latitudes and longitudes. Also included therein are two tables of distances in miles between Ahmedabad and Rajkot, and between Agra and Trichinopoly, with many intermediary cities, together with two pages of 'Dedication', one in Marathi language, and the other its English rendering. On the title cover and the inside page, the title of the atlas is given in both English and Marathi.

The publication by litho-process of a world atlas in Marathi language in 1836 in Bombay points to the fact that the then prevailing cartographic printing techniques and the geographical knowledge of the then known world were very well grasped by the professionals at that time in India.

Another point worth noting is that the text portion of the atlas contains only the definitions of elementary geographical terms like Cape, Isthmus, Ocean, Sea, Gulf, Strait, Hemisphere, Doldrums, Artic region, Antartic region, Tropical Zones, Sub-tropical zones, Latitude, Longitude, etc., indicating that the geographical knowledge was in its elementary stage in India in the early nineteenth century.

The atlas was published in Marathi, a regional language written and spoken in an area of which a few early topographical-like maps have been discovered in the recent past. These maps look more akin to properly drawn utilitarian maps, as discussed earlier. This happened in that part of India which could boast of a well organised contemporary naval fleet, commercial as well as for the purposes of war. This does not seem to be just a coincidence. It is more likely that the Marathas had seen and understood the utility of both sea charts and terrestrial maps, hence their interest in world geography.

All these factors, together with the sentiments expressed in the 'Dedication' of the Atlas *to diffuse Geographical knowledge among the Natives of India* (emphasis added), clearly show that there was not much indigenous knowledge of terrestrial map-making in early nineteenth century in India. Nowhere in the 'atlas' is there even a vague hint of any early indigenous mapping activity, or interest in map-making as a subject of study among the early Indians. The assertion of publishers in the 'Preface' of the Atlas that their venture was the *first ever attempt in the country* (emphasis added), seems to be very significant.

From the scrutiny of early Indian indigenous terrestrial mapping we may now pass on to the examination of the early navigational charts of Indian origin that have had no foreign influence.

As we know, Indian maritime activities go back to antiquity. The Indians had been exporting timber and other items to ancient Babylonia as early as 3000 BC. Much before the European navigational movement towards the East, by coasting all the way, Indians had mastered the art of cutting across the Indians Ocean and were navigating from the western Indian ports direct to southern Arabia and down the east African coast. We have also noted the references made earlier by writers to the existence of Indian indigenous sea charts.

Prof. B. Arunachalam of the Bombay University in his paper 'The Heaven-Finding Art in Indian Navigational Traditions and Cartography'[25] presented at the International Seminar on the Indian Ocean at Delhi in 1985, has referred to Indian sea charts and the existence of two separate manuals in Kutchi dialect of Gujarat, dated 1664[26] and 1780.[27] The first manual was discovered among the records of the National Archives of India and the other was acquired by the National Museum.

The 1664 manual, apart from written text covering the area, contains a set of sea charts in five parts. Together, these cover the region south of the latitude of Coondapur on the west coast of the peninsula, corresponding to modern Karnataka, Kerala and the southern Tamil Nadu coasts. It contains an inset map of Sri Lanka as well. The other manual of 1780 explains 'how the sailing-route between Madwad on the Saurashtra shore... and Daman could be charted with the help of Ardra (Orimionis) and *Chitra* (Spica Virginis)....' Stellar constellations. Prof. Arunachalam has studied these two sets of sea charts in some detail and has proved beyond any doubt the Indian origin of these map sheets as well as their utility to early Indian sailors, explaining the way the charts were being interpreted and made use of. These sea charts seem to be copies of the original charts in use then, or these could be the original charts in use in those days.

Another Kutchi map that has appeared on the scene and has a great bearing on the sea charts, discussed above, was the one acquired by Alexander Burnes[28] (1805-41) while 'sailing from Bombay to Kutch in May 1835' on the boat of a Kutchi sailor. Burnes stated that the boat had just completed the voyage of the Red Sea and part of the coast of Arabia. While on boat, he acquired the sea chart from the 'moalum' or the pilot of the boat.[29]

According to Burnes, the pilot had shown him some copies of the charts of English navigators. These were brought from Holland about a century ago by one Ram Singh. He is further reported to have said that he found on examination that the copies and the chart he had acquired, were not similar in nature or even in appearance.[30] Burnes also mentioned that the English charts belonged to eighteenth century and were prepared by John and Samuel Thornton. The acquired charts covered the area of the Red Sea from Jedda to the Horn of Africa. At the south end of the Red Sea the areas covered by the Gulf of Aden and the island of Socotra are also shown. Prof. Arunachalam compared Burnes' map by procuring a copy of it from the Paris archives where the original was lodged, with the other Kutchi maps in his possession. He concluded that the chart from Paris was similar to those he already had, i.e. the chart in Kutchi script of Gujarat. Its style and symbolism along with the entire technical data including polar altitudes, Stellar bearings and distance measurements were identically represented. Arunachalam had also conducted the final test by making a map-transformation by using the techniques and methodology employed in earlier Kutchi maps in his custody, and data given on the Paris chart. The result was quite successful; the map, thus prepared, corresponded to the modern map of the region.

A. Brunes, after procuring the said chart, had written its title on the chart along with the information he got from the Kutchi Pilot about the map, in his own hand. He wrote 'A native Indian chart of the coast of Arabia and the Red Sea drawn by an inhabitant of Cutch and used by pilots at the present time in that navigation given to me by a pilot.'[31] He had also rendered the place names in English which were known to him or could be identified by him in his own hand.

Such Kutchi navigational charts and the English maps of Thornton[32] differ in three main aspects. In the English charts, the compass rhumb lines are marked. But the Kutchi charts, instead, contain the relevant Stellar rhumb lines only. The coastline on the earlier charts is depicted more or less in its detailed form, but on the later charts it is shown as a straight line, almost parallel to its Stellar rhumb line. The mountain ranges on the land appear in their elevation forms on the English charts. The Kutchi charts, on the other hand, show the skyline profile of the ridges, sketched in rough hand, as seen from a particular point of entry to the coastal waters.[33]

Generally speaking the Kutchi charts contain a limited amount of information, just enough for identifying the location. Only islands, reefs, shoals, and banks, lights, port and other landmark identities, pole-star latitude coordinates, depth of water and the like, i.e. all the features of utility from seamen's point of view are shown. The land area, likewise, is shown without many details, 'barring coastal data and skyline profiles'. Such charts are not considered 'all purpose maps', but 'relevant only to the sailor on board'. These include only such information as would be needed for locating one's own position in coastal waters, including the distance indications, in terms of sailing time measured from the observers, position in the open sea to the conspicuous landmarks by which individual landing points are recognised.[34]

Prof. Arunachalam strongly believes that there is a distinct possibility of locating more of these Kutchi sea charts from the old seafaring families living on the western coast of India, in general and in Cutch-Kathiawar-Cambay-Gujarat coastal area above the Tapti river in particular. This was, most probably, the region from where the Aryans, or Harappan people of the Indus valley civilization started out on their maiden coastal voyages towards the west and reached Babylonia in the remote past. Or pioneering voyages were perhaps made by the original inhabitants of the southern India, the Dravadians. This is a point to be further investigated. Their early voyages, first coastal and later directly cutting across the Indian Ocean for centuries, gave them enough experience and consistency in the use of naval charts, which, by and by, must have shown steady improvement in their contents and designs with the increase of the frequency of their contacts with the Westerners of various nationalities.

When the Indians lost control of the overseas trade routes, the frequency of the voyages also decreased proportionally. They had no motivation left for long voyages to other seas once their own political power on the main land declined and the foreigners gained political and military ascendancy in India.

By the middle of the nineteenth century almost all the native powers had been subdued. No Indian dared to act in major commercial, military and political ventures without taking into consideration the views of their overlords, and we may even surmise, they could not move independently without obtaining concurrence from the English. Where then was the need, opportunity or incentive for native 'cartographers' to prepare or improve upon their earlier maps? The process of disinterest and lack of incentives to the native navigators and loss of commercial gains to the native rulers, began. The process first started with the rulers in the southern India before it spread to the entire country. The need for a topographical map for the native rulers of the Carnatic was over by the middle of the eighteenth century. The uprising in 1857, was, perhaps, the last chance when the Indian rulers and military generals could use their own maps, compiled and drawn by them, if at all they were conversant with their use.

If the old sea charts were prepared on some suitable and lasting material, like parchment, there is every possibility of some more of them lying buried in the personal or family archives of the descendants of the rulers of the forts and of the wealthy and leading commercial giants who were then controlling the maritime trade. Such charts could even be found in the archives of the famous temples on the sea shore, like Dwarka, Vijaygarh, Janjira, etc.

It is only natural that while on voyages, any new and important landmark seen by the sailors in those early days, would surely have been either mentally remembered or even recorded by them. If the possibility of this human activity is accepted, one can visualise that the position of the then Indian sea charts was almost similar to the early European land maps of the countries and areas then being newly discovered by the early travellers and constantly being improved upon by adding new information. Since the art of map-making was very expensive, only the rulers or rich people, mostly those engaged in trade and commerce, could afford to get topographical and route maps commissioned. When these very persons lost their positions to the English, incentives for the indigenous map-makers were also gone. Another important factor in the probable decline of the Indian map-making was that the major commercial, political and military decisions gradually came under the control of the new rulers who trusted their own maps and probably never encouraged the native map-makers.

European map-making could progress and flourish because of the urge of the people for travel, adventure and discovery of new lands, initially for commercial gains, and later for colonisation, and also because of the demand from a literate public. There were corresponding facilities available in their own countries for map-printing. There were also official requirements of the governments and rulers who also encouraged these activities. There were, however, no such demand or facilities in India. There were no printing presses, and manuscript hand-copying was very expensive which only the enlightened, the educated and the few rich could afford. The general public was mostly uneducated and unlettered. Those who could read and write were, by the force of tradition and circumstances, given mostly to religious practices and philosophical thoughts. There was no tradition either in any of the Indian languages, of illustrating with charts, the few travel stories that were hand-written for the small number of readers.

The cartography, as generally understood, is the science and art of map-making. A map is made for a purpose. This purpose entitles a pictorial representation of earth to be called a map. It is this aspect which differentiates a landscape painting from a topographical map. We have to dig out, though it would be extremely difficult, the probable purpose behind the creation of such 'map-painting' to make it eligible for consideration as a cartographic map. It was the absence of this 'purpose' from the Indian scene in early days which accounted for the absence of an Indian equivalent to the word 'map'. Is it just a landscape painting for painting's sake, painted by an artist for himself, or made to order? If there is a road depicted in the landscape, can it be termed a 'route map'? A part of the earth in the shape of grassy plain, a part of a river and a hill or a mountain with trees shown in a painting, do these topographical elements being represented there, entitle the painting to be considered a cartographic representation or a map? We have to be very careful while looking for answers to such questions.

Another point worth noting is that till the early eighteenth century maps used to be the product of individual manual work which included a lot of artistry. These were gradually 'stripped of their artistic decoration and transformed into plain specialised sources of information based upon measurement'. By the second half of the eighteenth century the individual 'craftsmanship was finally superseded by specialised science and the machine'.[35]

Let us now analyse the overall view of the entire scenario so far unveiled. Map-making, even in early days, was never a casual indulgence of an artist or a map-maker, but was a serious business of drawing, with a scientific approach to directions and distances, even though the data might have been of approximate nature.

One must also ponder over the question of how and why maps were produced and kept on developing in other countries? Once made, what made them grow year after year, century after century? If there had been nothing to sustain the activity, if there had been no need of maps, it would surely have died its natural death.

Even assuming that painting in early days was synonymous with map-making, then why was no progress made in the quality and contents of such 'map-making' for centuries? In spite of tremendous strides in the field of cosmology, astronomy and geographical knowledge, why did nobody think of utilising these developments in the art of 'map-making', even when Indians were in continuous contact with the Western civilization since 3000 BC.? Obviously, the assumption is erronous. Painting was painting, and is different from map-making.

Another curious aspect of the Indian indigenous find is that all the 'map-illustrations', both in traditional two-dimensional painting-forms, like 'religious map-illustrations', described earlier, as well as somewhat realistic cartographic representations like Maratha country side maps covering large areas, almost all belong to the eighteenth and early nineteenth centuries. A very few of them, that too sea charts only, are of a period earlier than eighteenth century. Even if useful in their own days, the utility of these rough, inaccurate, imaginative, unrealistic, and unscientific Indian indigenous 'map-paintings' or 'illustrations' disappeared altogether by the later half of the nineteenth century. The map-making in India today has no relation whatsoever to this earlier tradition.

Yet there is still some hope of discovering some more sea charts, giving a semblance of their continual growth.

On the other hand, discovery of more land maps of different periods and from different regions may, by comparison, give us the answers to some of the perplexing questions in future. The possibility of this, however, is very remote. Under the circumstances, the story of Indian indigenous maps would always be somewhat incomplete.

NOTES

1. D.C. Sircar, op. cit., p. 60.
2. Ibid., pp. 60-3.
3. R.A.L.H. Gunawardana, op. cit., pp. 54-89.
4. D.V. Kale, 'Maps and Charts in Bharata Itihasa Samsodhaka Mandla', in *Bharata Itihasa Samsodhaka Mandala Quarterly,* Pune, October-December 1948, pp. 60-5.
5. Ibid., pp. 62-3.
6. Ibid., p. 64.
7. Ibid., p. 62.
8. D.C. Sircar, op. cit., pp. 60-3.
9. S. Ray (ed.), *Uttarasamacharita,* Calcutta, 1934, p. 106, quoted by D.C. Sircar, op. cit., p. 60.
10. R.H. Phillimore, 'Three Indian Maps', in *Imago Mundi,* Amsterdam, Vol. IX, 1952, pp. 111-14.
11. S.N. Prasad, op. cit., Nos. F. 9/10-11.
12. Ibid., No. F.124/2.
13. R.H. Phillimore, 'Indian Maps', op. cit., p. 113.
14. James Rennell, op. cit., p. 300.
15. R.H. Phillimore, 'Indian Maps', op. cit., p. 112.

16. Ibid., p. 114.
17. C.D. Deshpande, 'A Note on Maratha Cartography', in *The Indian Archives*, New Delhi, 1953, Vol. VII, pp. 87-94.
18. Susan Gole; *Indian Maps,* op. cit., Map No. 69, p. 145.
19. Irfan Habib, 'Cartography in Mughal India', in *The Indian Archives,* New Delhi, 1980, Vol. No. XXVIII, pp. 68-105; re-printed from *Medieval India—A Miscellany,* Vol. IV.
20. Ibid., pp. 88-9.
21. Susan Gole has used the main caste title of 'kayath', for 'Kayasth', instead of the sub-caste, i.e. Saksena.
22. Susan Gole, *Indian Maps,* op. cit., p. 91.
23. Ibid., pp. 120-5.
24. Pandurung, Dadoba and Nana Narayun, *Marathee Atlas containing nine Maps,* Bombay, 1836.
25. B. Arunachalam, op. cit., pp. 191-221.
26. Ibid., p. 203.
27. Ibid., p. 195.
28. B. Arunachalam, op. cit., p. 218; Susan Gole, *Indian Maps,* op. cit., pp. 156-7.
29. Susan Gole, *Indian Maps,* op. cit., p. 156.
30. Ibid., p. 156.
31. B. Arunachalam, op. cit., p. 219, Footnote No. 42.
32. Susan Gole, *Indian Maps,* op. cit., p. 156; *Early Printed Maps,* op. cit., Map No. 22.
33. B. Arunachalam, op. cit., p. 205.
34. Ibid., p. 205.
35. Leo Bagrow, op. cit., pp. 21-2.

Bibliography

PART I

The survey of India, being a subordinate department since its inception in 1767, was always under the authorities of one department of the central government or the other. The central government was, in turn, being controlled earlier by the Court of Directors during the East India Company period and later by the British government through the Secretary of State for India.

The relevent original MS/Printed records of the following departments/ministry dealing with practically every aspect of the Survey of India, lodged with the National Archives of India were examined:

1. Military Department ....till 1818
2. General Department 1818-23
3. Military Department 1823-65
4. Home Department 1865-71
5. Revenue, Agriculture & Commerce Department 1871-9
6. Home, Revenue & Agriculture Department 1879-81
7. Revenue & Agriculture Department 1881-1923
8. Education, Health & Lands Department 1923-45
9. Department/Ministry of Agriculture 1945-51
10. The Survey of India [including MS map records (1700-1900) and the printed Annual Reports] 1700-1956

PART II

Arunachalam, B., 'The Heaven-Finding Art in Indian Navigational Traditions and Cartography,' in Satish Chandra, (Ed.), *The Indian Ocean Explorations in History, Commerce and Politics*, New Delhi, 1987.

Bagrow, Leo, *History of Cartography*, Eng. edn., revised and enlarged by R.K. Skelton, London, 1964.

Black, Charles E.D., *A Memoir on the Indian Surveys, 1875-1890*, London, 1891.

Cambridge, R.O., *An Account of the War in India between the English and the French on the Coast of Choromandel ... from 1750 to 1760*, London, 1761.

Campos, J.J.A., *History of the Portuguese in Bengal*, Calcutta, 1919.

Coldstream, W.M., *Records of the Survey of India, Vol. XII, Notes on Surveys of India Maps*, Calcutta, 1919.

Cook, A.S., 'Major James Rennell and a Bengal Atlas (1780 and 1781),' in *Foreign and Commonwealth Office Library and Records Report*, London, 1976.

——, 'Maps from a Survey Archive: India Office Collection' in The British Library, *The Map Collector*, London, Sept. 1984.

Cunningham, A., *Geography of Ancient India*, London, 1871.

Deshpande, C.D., A Note on Maratha Cartography, in *The Indian Archives*, Vol. VII, New Delhi, 1953.

Enbury, E., *History of Ancient Geography*, 2nd. Edn., Vol. I, London, 1883.

Encyclopedia Britannica, 1967 (edn.).

Gole, Susan, *Early Maps of India*, New Delhi, 1976.

——, *India Within the Ganges*, New Delhi, 1983.

——, *A Series of Early Printed Maps of India in Facsimile*, New Delhi, 1984.

——, *Indian Maps and Plans from the earliest times to the advent of European Surveys*, New Delhi, 1989.

Gunawardana, R.A.L.H.,'Changing Patterns of Navigation in the Indian Ocean and their Impact on Precolonial Sri Lanka', in Satish Chandra (ed.), *The Indian Ocean Explorations in History, Commerce and Politics*, New Delhi, 1987.

Hirst, F.C., *The Surveys of Bengal by James Rennell, 1764-1777*, Calcutta, 1917.

Hourani, G.F., *Arab Seafaring in the Indian Ocean in Ancient and Early Medieval Times*, Princeton, N.J., 1951.

Imperial Gazetteer of India, Calcutta, 1908 (edn.).

Imperial Library, *Maps and Plans*, Calcutta, 1910.

India Office, *A Catalogue of Manuscript and Printed Reports, Field-Books, Memoirs, Maps, etc., of the Indian Survey Deposited in the Map Room of the India Office*, London, 1878.

Irfan Habib, 'Cartography in Mughal India', in *The Indian Archives*, Vol. XXVIII, New Delhi, 1980.

Janaki, V.A., 'Gujarat as the Arab knew it', in Maharaja Sayajirao University of Baroda, *Geography Research Paper Series*, No. 4, Baroda, 1969.

Kale, Prof. D.V., 'Maps and Charts in Bharata Ithiasa Samsodhak Mandala', in *Bharata Itihasa Samsodhak Mandala Quarterly*, Poona, 1948.

Kalota, N.S., *India as Described by Megasthenes*, Delhi, 1978.

Lach, Donald F., *Asia in the Making of Europe*, 2 vols., Chicago and London, 1965.

Low, C.R., *History of Indian Navy*, 2 Vols., London, 1877.

Madan, P.L., 'Cartographic Records in the National Archives of India', in *Imago Mundi*, Amsterdam, 1971.

Malcolm, John, *Life of Robert, Lord Clive*, Vol. II, 1836.

Markham, Clement R., *Memoir on the Indian Surveys*, 2nd edn., 1878.

McCrindle, J.W., *Ancient India as Described in Classical Literature*, Westminster, 1901.

Misra, R.P., *Fundamentals of Cartography*, Mysore, 1969.

Mukherjee, R.K., *A History of Indian Shipping*, 2nd. edn., 1957.

Nag, Prithvish, 'The Indian Ocean: India and Africa: Historical and Geographical Perspectives', in Satish Chandra (ed.), *The Indian Ocean Explorations in History, Commerce and Politics*, New Delhi, 1987.

Pandurung, Dadoba and Nana Narayan, *Marathee Atlas containing nine maps*, Bombay, 1836.

Parry, J.H., *The Age of Reconnaissance*, London, 1963.

Phillimore, R.H., *Historical Records of the Survey of India*, 5 vols., Dehra Dun, Vol. I, 1945, Vol. IV, 1958.

——, 'Three Indian Maps', in *Imago Mundi*, Vol. IX, Amsterdam, 1952.

Prasad, Dr. S.N.(ed.), *Catalogue of the Historical Maps of the Survey of India, 1700-1900*, New Delhi, 1975.

Rawlison, H.G., 'Early Contacts between India and Europe', in A.L. Basham (ed.), *A Cultural Heritage of India*, Oxford, 1975.

Ray, S. (ed.), *Uttarasamacharita*, Calcutta, 1934.

Rennell, James, *Memoir of a Map of Hindoostan*, 3rd edn., London, 1793.

Sastri, K.A.N., *History of South India from Pre-historic Times to the Fall of Vijayanagar*, Madras, 1955.

Sastri, S.M. (ed.), *M. Crindle's Ancient India as described by Ptolemy*, Calcutta, 1927.

Satish Chandra (ed.), *The Indian Ocean Explorations in History, Commerce and Politics*, New Delhi, 1987.

Sebock, Lou (ed.), Public Archives of Canada, National Map Collection, *Atlases Published in the Netherlands in the Rare Atlas Collection*, Provisional Series, No.1, Ottawa, 1973.

Sircar, D.C., *Geography of Ancient and Medieval India*, Varanasi, 1950.

Tirmizi, S.A.I. (ed.), *Catalogue of the MRIO — Miscellaneous Maps of the Survey Of India, 1742-1872*, New Delhi, 1982.

Tooley, R.V., *Map and Map-Makers*, London, 1970.

Verlinden, Charles, 'The Indian Ocean: The Ancient Period and the Middle Ages', in Satish Chandra (ed.) *The Indian Ocean Explorations in History, Commerce and Politics*, New Delhi, 1987.

Wallis, Helen, 'Missionary Cartographers to China', in *Geographical Magazine*, Vol. 47, No. 12, London, 1975.

Xinhua, Chinese daily News-paper, quoted in *Indian Express*, New Delhi, 18 May, 1989.